OPPORTUNITIES IN HIGH MAGNETIC FIELD SCIENCE

Committee on Opportunities in High Magnetic Field Science

Solid State Sciences Committee

Board on Physics and Astronomy

Division on Engineering and Physical Sciences

NATIONAL RESEARCH COUNCIL
OF THE NATIONAL ACADEMIES

THE NATIONAL ACADEMIES PRESS
Washington, D.C.
www.nap.edu

THE NATIONAL ACADEMIES PRESS • 500 Fifth Street, N.W. • Washington, DC 20001

NOTICE: The project that is the subject of this report was approved by the Governing Board of the National Research Council, whose members are drawn from the councils of the National Academy of Sciences, the National Academy of Engineering, and the Institute of Medicine. The members of the committee responsible for the report were chosen for their special competences and with regard for appropriate balance.

This study was supported by Grant No. DMR-0318562 between the National Academy of Sciences and the National Science Foundation. Any opinions, findings, conclusions, or recommendations expressed in this publication are those of the author(s) and do not necessarily reflect the views of the organizations or agencies that provided support for the project.

International Standard Book Number 0-309-09582-4 (Book)
International Standard Book Number 0-309-54847-0 (PDF)
Library of Congress Control Number: 2005924159

Cover (clockwise from left): Axial MRI scan of the human brain; decoherence of electron spins near a quantum phase transition by a nuclear spin bath, imaged by inelastic neutron scattering as a function of magnetic field, courtesy H.M. Ronnow, Commissariat à l'Energie Atomique at Grenoble, France; Meissner-effect levitation of a high-temperature superconductor over a rare-earth magnet, courtesy F. Kraehenbuehl, Cables Cortaillod S.A.; ribbon representation of a large protein complex composed of two copies of the *E. coli* histidine phosphocarrier protein (in green) and two copies of subunit A of the enzyme II mannose protein (in red and blue), courtesy M. Clore and D.C. Williams, National Institutes of Health.

Additional copies of this report are available from the National Academies Press, 500 Fifth Street, N.W., Lockbox 285, Washington, DC 20055; (800) 624-6242 or (202) 334-3313 (in the Washington metropolitan area); Internet, http://www.nap.edu

Printed in the United States of America

THE NATIONAL ACADEMIES

Advisers to the Nation on Science, Engineering, and Medicine

The **National Academy of Sciences** is a private, nonprofit, self-perpetuating society of distinguished scholars engaged in scientific and engineering research, dedicated to the furtherance of science and technology and to their use for the general welfare. Upon the authority of the charter granted to it by the Congress in 1863, the Academy has a mandate that requires it to advise the federal government on scientific and technical matters. Dr. Bruce M. Alberts is president of the National Academy of Sciences.

The **National Academy of Engineering** was established in 1964, under the charter of the National Academy of Sciences, as a parallel organization of outstanding engineers. It is autonomous in its administration and in the selection of its members, sharing with the National Academy of Sciences the responsibility for advising the federal government. The National Academy of Engineering also sponsors engineering programs aimed at meeting national needs, encourages education and research, and recognizes the superior achievements of engineers. Dr. Wm. A. Wulf is president of the National Academy of Engineering.

The **Institute of Medicine** was established in 1970 by the National Academy of Sciences to secure the services of eminent members of appropriate professions in the examination of policy matters pertaining to the health of the public. The Institute acts under the responsibility given to the National Academy of Sciences by its congressional charter to be an adviser to the federal government and, upon its own initiative, to identify issues of medical care, research, and education. Dr. Harvey V. Fineberg is president of the Institute of Medicine.

The **National Research Council** was organized by the National Academy of Sciences in 1916 to associate the broad community of science and technology with the Academy's purposes of furthering knowledge and advising the federal government. Functioning in accordance with general policies determined by the Academy, the Council has become the principal operating agency of both the National Academy of Sciences and the National Academy of Engineering in providing services to the government, the public, and the scientific and engineering communities. The Council is administered jointly by both Academies and the Institute of Medicine. Dr. Bruce M. Alberts and Dr. Wm. A. Wulf are chair and vice chair, respectively, of the National Research Council.

www.national-academies.org

1939-2004

The Committee on Opportunities in High Magnetic Field Science dedicates this report to a dear friend and valued colleague, Jack E. Crow. His vision, enthusiasm, and energy helped to move high magnetic field research forward, and his strong voice helped bring it to the attention of the nation. The committee was honored to hear from Jack at its December 2003 meeting at the National High Magnetic Field Laboratory; his words were as wise, patient, and humorous as always. Photo courtesy of the National High Magnetic Field Laboratory.

COMMITTEE ON OPPORTUNITIES IN HIGH MAGNETIC FIELD SCIENCE

PETER B. MOORE, Yale University, *Chair*
GABRIEL AEPPLI, University College London
MEIGAN ARONSON, University of Michigan
PAUL M. CHAIKIN, Princeton University
PAUL D. ELLIS, Pacific Northwest National Laboratory
PETER F. GREEN, University of Texas at Austin
DAVID C. LARBALESTIER, University of Wisconsin at Madison
J. DAVID LITSTER, Massachusetts Institute of Technology
JOSEPH MINERVINI, Massachusetts Institute of Technology
J. MICHAEL ROWE, National Institute of Standards and Technology
JOHN M. ROWELL, Arizona State University
MANSOUR SHAYEGAN, Princeton University
ROBERT TYCKO, National Institutes of Health
VALERII VINOKUR, Argonne National Laboratory

Staff

DONALD C. SHAPERO, Director
TIMOTHY I. MEYER, Program Officer
MAUREEN MELLODY, Program Officer (August 2003-December 2003)
DAVID B. LANG, Research Assistant
PAMELA A. LEWIS, Program Associate
VAN AN, Financial Associate

SOLID STATE SCIENCES COMMITTEE

MARC A. KASTNER, Massachusetts Institute of Technology, *Chair*
PETER F. GREEN, University of Texas at Austin, *Vice Chair*
DAVID D. AWSCHALOM, University of California at Santa Barbara
ANGELA M. BELCHER, Massachusetts Institute of Technology
ROBERT J. CAVA, Princeton University
JOHN CLARKE, University of California at Berkeley
DUANE B. DIMOS, Sandia National Laboratories
JAMES P. EISENSTEIN, California Institute of Technology
PETER C. EKLUND, Pennsylvania State University
PATRICK D. GALLAGHER, National Institute of Standards and Technology, Center for Neutron Research
SHARON C. GLOTZER, University of Michigan
BARBARA JONES, IBM Almaden Research Center
STEVEN A. KIVELSON, Stanford University
HERWIG KOGELNIK, Bell Laboratories and Lucent Technologies, Inc.
ANTHONY J. LEGGETT, University of Illinois at Urbana-Champaign
M. BRIAN MAPLE, University of California at San Diego
SIDNEY R. NAGEL, University of Chicago
ARTHUR P. RAMIREZ, Lucent Technologies, Inc.
A. DOUGLAS STONE, Yale University
CHRIS G. VAN DE WALLE, University of California at Santa Barbara

Staff

DONALD C. SHAPERO, Director
TIMOTHY I. MEYER, Program Officer
DAVID B. LANG, Research Assistant
VAN AN, Financial Associate

BOARD ON PHYSICS AND ASTRONOMY

Acknowledgment of Reviewers

This report has been reviewed in draft form by individuals chosen for their diverse perspectives and technical expertise, in accordance with procedures approved by the National Research Council's Report Review Committee. The purpose of this independent review is to provide candid and critical comments that will assist the institution in making its published report as sound as possible and to ensure that the report meets institutional standards for objectivity, evidence, and responsiveness to the study charge. The review comments and draft manuscript remain confidential to protect the integrity of the deliberative process. We wish to thank the following individuals for their review of this report:

Elihu Abrahams, Rutgers University,
Jack Bass, Michigan State University,
W.F. Brinkman, Princeton University,
Frances Hellman, University of California at Berkeley,
Franz Himpsel, University of Wisconsin at Madison,
Gottfried Landwehr, University of Wuerzburg,
M. Brian Maple, University of California at San Diego,
Peter Schiffer, Pennsylvania State University,
Gerhard Wagner, Harvard Medical School, and
Albert Zeller, National Superconducting Cyclotron Laboratory.

We also wish to thank the following individuals for their review of the committee's interim letter report:

Shirley Chiang, University of California at Davis,
Linda J. (Lee) Magid, University of Tennessee at Knoxville,
Thomas Mareci, University of Florida, and
Peter Wanderer, Brookhaven National Laboratory.

Although the reviewers listed above have provided many constructive comments and suggestions, they were not asked to endorse the conclusions or recommendations, nor did they see the final draft of the report before its release. The review of this report was overseen by Mildred S. Dresselhaus, Massachusetts Institute of Technology. Appointed by the National Research Council, she was responsible for making certain that an independent examination of this report was carried out in accordance with institutional procedures and that all review comments were carefully considered. Responsibility for the final content of this report rests entirely with the authoring committee and the institution.

Acknowledgments

The members of the Committee on Opportunities in High Magnetic Field Science wish to thank the nonmembers who made formal presentations at its meetings (their names appear in Appendix D). Their presentations and the ensuing discussions were extremely informative and had a major impact on the committee's deliberations. The committee also thanks those who sent in letters and e-mail messages in response to its public request for input from the very large community of scientists who use high-field magnets. The committee is particularly grateful to the National High Magnetic Field Laboratory in Tallahassee both for its hospitality when the committee met there and for its willingness to discuss with the committee every aspect of its operations.

It would be impossible for the members of a National Research Council committee to produce a useful report without the help of NRC staff. Timothy Meyer and Donald Shapero of the Board on Physics and Astronomy guided us through the entire process. Their wise advice helped shape our report, and their hard work ensured its timely production.

Contents

Executive Summary

In response to an informal request from the National Science Foundation, the National Research Council convened the Committee on Opportunities in High Magnetic Field Science in mid-2003. The committee was charged with four tasks:

- Assessment of the current state and future prospects of high-field magnet science and technology in the United States.
- Assessment of the status of U.S. high-field efforts in the international context, and trends in the international arena.
- Identification of particularly promising multidisciplinary areas for research and development with respect to magnetic fields.
- Discussion and prioritization of any major new initiatives in the construction of high-field magnets for the coming decade.

The committee focused its attention on identifying the compelling scientific opportunities the field affords and the institutional infrastructure that would be required to realize them. Its conclusions and recommendations follow.

A magnet is "high field" if its field strength is great enough to test the limits of the mechanical and/or the electromagnetic properties of the materials from which it is built. High-field magnets have been—and continue to be—used for research in many scientific disciplines, including medicine, chemistry, and condensed-matter physics; they are also enabling for fields such as plasma science and high-energy physics. Research that could only have been done with such magnets has produced important insights in a host of areas, ranging from brain function to high-temperature superconductivity. High magnetic fields are of great interest in areas such as astrophysics and magnetohydrodynamics, and high-field magnets also play an increasingly important role in industry.

The committee's task was to identify key scientific and technological challenges and opportunities, not to make specific programmatic recommendations. In general, the committee found that high magnetic field science in the United States is healthy and broadly multidisciplinary. However, there are some important opportunities that will be missed unless attention is paid to them soon.

Conclusion. *High magnetic field science and technology are thriving in the United States today, and the prospects are bright for future gains from high-field research.*

Recent accomplishments include the development of functional magnetic resonance imaging (MRI), which is revolutionizing neuroscience; optically pumped magnetic resonance techniques, which allow visualization of new quantum phenomena in semiconductors; and ion cyclotron resonance mass spectroscopy, which is becoming an important tool for exploring the chemical composition of complex systems. High-field research has led to the discovery of new states of matter in low-dimensional systems. It has also provided the first indications of how high-temperature superconductors evolve into unconventional metallic alloys in the extreme quantum limit. Outstanding work continues to be done in the area of magnet engineering, the discipline on which all these activities depend. There is every reason to believe that there will be new accomplishments as interesting as those mentioned above in the decades to come, especially if magnets are built that deliver higher fields than those available today. For instance, pulsed fields offer the opportunity to explore the highest magnetic fields in ways that can take research in new directions. Additionally, advances in high-speed electronics, instrumentation, and miniaturization could also allow greater experimental access to higher fields.

Conclusion. *The United States is a leader in many areas of high-field science and technology, but further investment will be required to make it competitive in some critical areas.*

There are many indicators of the strength of the U.S. effort in high magnetic field research. For example, condensed-matter physicists and materials researchers from other parts of the world routinely travel to the National High Magnetic Field Laboratory (NHMFL) to perform experiments that they are unable to do at home, but U.S. scientists seldom travel abroad for that reason. An important corroborating observation can be found in the European Science Foundation's 1998 report *The Scientific Case for a European Laboratory for 100 Tesla Science*, which states that one of the prime motivations for such a facility was "to be competitive with laboratories elsewhere, particularly in the United States and Japan."[1] In addition, the superconducting magnets being installed in the Large Hadron Collider (LHC) at the European Organization for Nuclear Research (CERN), as well as those contemplated for the International Thermonuclear Experimental Reactor (ITER), depend on magnet technology developed in the United States (although the magnets are in fact being manufactured overseas), as do the magnets installed in several other user facilities overseas.

By way of contrast, in the area of nuclear magnetic resonance (NMR), which is an important component of high-field science, the United States is competitive but not dominant. About half of the instrumentation used by NMR spectroscopists in the United States, and virtually all of the magnets in their spectrometers, were manufactured abroad. Further, many of the most important recent advances in NMR have been made overseas, and, in general, European and Japanese companies have been ahead of U.S. companies in commercializing magnet technology advances. Finally, Europe is far ahead of the United States in equipping its synchrotron light sources and neutron scattering centers with instruments for studying the x-ray- and neutron-scattering properties of materials in high magnetic fields. It also worth noting that several key facilities in Japan have made important contributions to the development of the technologies required for the generation of the highest steady-state and pulsed magnetic fields.

Conclusion. *High-field magnet science is intrinsically multidisciplinary.*

The construction of high-field magnets has always been motivated by the science that could be done with them, and in recent decades, physics, chemistry, biology, and medicine have all benefited from advances in magnet technology. Even the technology of high-field magnets is cross-disciplinary. Materials science and engineering make dominant contributions, but several branches of physics contribute as well.

[1]European Science Foundation, *The Scientific Case for a European Laboratory for 100 Tesla Science*, ESF Studies on Large Research Facilities in Europe, 1998. Available online at http://www.esf.org/publication/109/100T.pdf.

Conclusion. *U.S. scientists will be unable to access a wealth of science opportunities if high magnetic field instrumentation is not provided at the Spallation Neutron Source and at the nation's third-generation light sources.*

The scientific opportunities that are available to those able to study the neutron and x-ray scattering properties of materials at high magnetic fields are attracting growing attention around the world. Certain aspects of magnetism and high-temperature superconductivity have already been elucidated overseas at scattering center laboratories that have high-field instrumentation. Nevertheless, the United States does not currently plan to increase the high magnetic field instrumentation at its national radiation laboratories. Unless steps are taken to rectify this situation, the United States is sure to lag behind in key areas of condensed-matter and materials physics.

Conclusion. *Some important issues relevant to the advancement of magnet technology could be more efficiently addressed if the interested constituencies would interact more strongly, communicate more fully, and coordinate their activities better.*

One striking characteristic of all the sciences that use high magnetic fields is how constrained they are by the limitations imposed by magnet technology. Nevertheless, despite a shared need to overcome the same set of fundamental problems, each constituency has historically tended to develop the magnets it needed without much reference to the others. The reasons are several and obvious. The communities that use high-field magnets have different missions, and the magnets they need are specific to the mission of each. In addition, these communities are supported by different funding agencies, each of which has had its own perspective. A coordinated approach to magnet technology based on the pooling of resources and talent would be beneficial.

Based on these conclusions, the committee has several recommendations that it offers, with the most important first.

Recommendation. *The United States should maintain a national laboratory that gives its scientific community access to magnets operating at the highest possible fields.*

A national high-field magnet facility is essential to the vitality of many important scientific disciplines. NHMFL has successfully fulfilled the need for high-field magnets for about a decade, and its activities have done much to foster the leadership position that the United States currently enjoys in many areas of magnetic science and technology. It is important to understand that at any high-field magnet laboratory, the capabilities of the devices available for controlling the environment in which a sample is tested and for measuring a sample's properties are almost as important as the field strengths of the magnets themselves. Thus it is vital that a national laboratory equip its magnets with the best possible supporting instrumentation and personnel. In addition, ways to maximize the return on capital invested in the national laboratory should be explored, such as longer hours of operation and flexible scheduling. The laboratory should undertake a cost-benefit analysis to identify the optimal balance between addressing user demand and the increased operating costs associated with longer hours of operation. For instance, the nation's synchrotron light sources and neutron-scattering centers provide access 24 hours a day, 7 days a week when in full operation; this schedule allows visiting researchers to use their time at the facility to the best advantage. The trade-offs for expanding access to the NHMFL need to be identified and weighed carefully, especially in constrained budget situations.

Recommendation. *New instruments for studying the neutron and x-ray scattering properties of materials in high magnetic fields should be developed in the United States.*

Nowhere in the domestic research program is the gap between the instrumentation available for experimentation at zero field and that available for high-field experimentation wider than in the areas of neutron and x-ray scattering. This gap in capability is unfortunate, because scattering experiments provide a powerful means for elucidating atomic and magnetic structure, as well as for determining the nature of the spatial and dynamical correlations in materials. Development of new high-field capabilities at x-ray- and neutron-scattering centers in the United States could have an enormous scientific impact.

Recommendation. *A consortium should be established to foster the development of magnet technology.*

Rather than supporting an all-out, brute-force effort to build higher-field magnets using current technology, it makes sense to find new approaches that will make it easier (and cheaper) to build the magnets needed for research. Essential to this enterprise will be the development of both resistive and superconducting materials with improved electrical, magnetic, and mechanical properties. Scientists and engineers from all the communities working today on magnet technology should be brought together: the magnet engineers at the NHMFL; academic researchers; the magnet designers in the high-energy physics and fusion communities; commercial vendors of superconducting magnets, including nuclear magnetic resonance and magnetic resonance imaging systems; and manufacturers of advanced materials, such as high-strength materials and superconducting wire.

The sharing of information and resources within the larger community, which is now fragmented into components that communicate poorly, would accelerate the rate at which solutions are found to the fundamental problems confronted by all. The committee proposes that the involved communities cooperate to establish a consortium for developing the technology necessary to pursue several aggressive goals that may have different timescales. Some groups might frame their goals in terms of application-specific requirements for magnet performance, such as the development of a 30-T superconducting high-resolution magnet for NMR, a 60-T DC hybrid magnet, or a 100-T long-pulse magnet. Others, such as the high-energy physics and fusion science communities, might focus explicitly on the materials problems intrinsic to enabling high-volume production of quality conductors for a variety of magnet systems.

> ***Recommendation.*** *Government agencies supporting high-field magnetic resonance research should directly support the development of technology and instrumentation for magnetic resonance and magnetic resonance imaging.*

Without the concomitant development of ancillary technologies, the construction of higher-field magnets for magnetic resonance will not produce the scientific dividends it should. While federal funding for the application of existing technology and methods to specific scientific problems has generally been good, federal funding for the development of novel technology and methodology has been poor. Magnetic resonance and MRI instrument manufacturers have done a good job of advancing the supporting technologies for these techniques when commercial markets for their products justified their doing so. However, there are many areas where technological advances are sorely needed but the commercial market is not large enough to attract the attention of instrument manufacturers. For example, optimal coils for high-field MRI will probably not be realized unless groups outside the commercial sector undertake a sizable research program. Likewise, because higher fields cause significant changes in the relative strengths of the interactions that determine how nuclear magnetic moments evolve, pulse sequences and methodologies will have to be improved if magnetic resonance research is to take full advantage of high-field magnet advances.

1

Introduction

THE IMPORTANCE OF MAGNETISM IN THE MODERN WORLD

Because humans do not sense magnetic fields, it took a long time for the importance of magnetism in the natural world to be appreciated. The first magnetic device to come into wide use was the magnetic compass. It is believed to have been invented in China around 200 B.C., but it was not fully understood until the 19th century, when systematic investigation of magnetic phenomena began. In 1819, H.C. Oersted discovered that electric currents engender magnetic fields, and a few years later, M. Faraday discovered how to use magnetism to interconvert mechanical and electrical energy—that is, how to build electric motors and dynamos. These advances and others related to magnetism had a profound impact on society.

Magnetic devices are now so deeply integrated into everyday life that the average citizen takes most of them for granted. It is assumed that when a light switch is turned on, a generator at some unknown location will deliver electric power to the appropriate lightbulb. It is assumed that when the ignition switch of an automobile is turned, an electric motor will start the car. It is assumed that when the appropriate sequence of key strokes is made on a computer keyboard, information will be recorded faithfully on a (magnetic) disc. About the only time people are likely to be aware of having encountered magnet technology is when they go to the hospital for a magnetic resonance imaging (MRI) scan. Even so, they are unlikely to realize that the drug used to treat the disease diagnosed by MRI is itself the product of research that relied heavily on magnet technology.

Magnetic devices are even more important in the scientific world than they are in everyday life. Instruments that take advantage of magnetic phenomena or that use magnets to produce electromagnetic radiation are used in many fields, the most spectacular being the accelerators employed in high-energy physics to study the structure of subatomic particles. High-field magnets are used to control particle trajectories in accelerators, and over the years, the interest of the high-energy physics community in increasing the energies at which accelerators operate has been a powerful driver of magnet technology.

THE SIGNIFICANCE OF HIGH MAGNETIC FIELD RESEARCH

Most of the magnetic devices important to the public do not generate high fields, MRI being the exception. Thus it is reasonable to ask why the nation should support high magnetic field science and technology. Before this question can be answered, some background information must be supplied. The committee starts by reminding the reader that there are two kinds of magnets: permanent magnets and electromagnets.

Permanent magnets are made of substances like iron (Fe), cobalt (Co), and nickel (Ni), the atoms of which have large magnetic moments. When those substances are in their unmagnetized states, the magnetic moments of their atoms are randomly oriented. Magnetization is achieved by making the magnetic moments of their constituent atoms point in the same direction, which can be done by exposing the substances to an external magnetic field. What distinguishes a magnetizable substance from another substance that contains atoms with magnetic moments is that once the magnetic moments of its atoms have become aligned, they remain that way. Thus, a piece of iron that has been exposed briefly to a magnetic field emerges with a net magnetic moment that persists; it has become a permanent magnet. A compass needle is a permanent magnet, and permanent magnets can produce fields up to about 2 T. Permanent magnets have many practical uses, and the search for magnetizable materials with improved properties is ongoing; its goals include increasing the efficiency of electrical motors and generators.

Electromagnets can be made of any material that conducts electricity, regardless of the magnetic properties of its atoms, and they produce magnetic fields via the Oersted effect whenever an electric current flows through them. Electromagnets are commonly made from coils of an electrical conductor. Since the field contributed by each turn in a coil adds to that of its neighbors, and the field per turn increases with electric current, the more turns in the coil and the greater the current put through it, the stronger the magnetic field that results. All high-field magnets—that is, magnets that generate fields substantially greater than 2 T (the limit of permanent magnetization for iron)—are electromagnets.

Scientists have been building electromagnets that deliver fields of ever-increasing strength since the 19th century. Two issues have had to be confronted at every step of the way. First, the field of an energized electromagnet exerts forces on its own structure that increase as the square of the field strength and that will destroy it if not contained. Second, if the electrical conductor of which the magnet is made is resistive, as was always the case before 1960 or so, heating may also cause it to fail. Thus, the construction of magnets that operate at high fields is, and has always been, an engineering challenge.

The justification for building ever more powerful magnets remains today what it was at the outset—namely, the scientific and, ultimately, the social benefits of research done using more powerful magnets. Over the years, advances in magnet technology have paid huge dividends, not only to those interested in the science made accessible by increases in field strength but also to the much larger community able to work at lower fields. The technological advances that have made it possible to build magnets that push the field strength envelope have often made it easier and cheaper to build magnets that produce less extreme fields.

There is every reason to believe that advances in magnet technology will be as rewarding in the future as they were in the past. Research on the magnetic properties of materials, particularly those that are superconducting (e.g., high-T_c and high-field superconductors, advanced sensors), and on magnet design and construction can have substantial economic payoffs. Electric motors and generators will operate with improved efficiency. More efficient ways will be found to transmit electric power over long distances. Better information storage devices will be made. On the scientific side, improvements in magnet technology will lead to better instrumentation for studying the structure and properties of materials of all kinds, at all length scales. The benefits to the public of most of the scientific advances will be less direct but no less real. It is appropriate, therefore, that the nation support high magnetic field research and that the state of high magnetic field science and technology in the United States be assessed periodically.[1]

THE TASK OF THE COMMITTEE

In the summer of 2003, the National Research Council established the Committee on Opportunities in High Magnetic Field Science in response to an informal request from the National Science Foundation (NSF). The committee was charged with (1) assessing the current state and future prospects of high-field magnetic

[1]See, for example, National Research Council, *High-Magnetic-Field Research and Facilities*, Washington, D.C., National Academy Press, 1979; and National Science Foundation, *Final Report of NSF Panel on Large Magnetic Fields*, Arlington, Va., National Science Foundation, 1988.

science and technology in the United States, (2) assessing the current status of U.S. high-field efforts in the international context, (3) identifying particularly promising multidisciplinary areas for research and development, and (4) reviewing major initiatives in the construction of new high-field magnets and setting priorities for the coming decade. In its deliberations, the committee took as its purview both the disciplines relevant to the generation of high magnetic fields and those that would benefit if higher fields could be generated.

DEFINITION OF HIGH MAGNETIC FIELD

The field a magnet produces is "high" if it tests the limits of the mechanical and/or electromagnetic properties of the materials of which the magnet is made. In many instances, the quantity that determines whether a magnet is high field is the amount of energy stored in its field, which is proportional to the integral of the square of its field strength over the volume affected. Thus, a magnet having a maximum field strength around 8 T and a bore large enough to accommodate a human being is as much a high-field magnet as the much smaller bore magnet in a nuclear magnetic resonance (NMR) spectrometer operating at 20 T. Some magnets operate in a pulsed mode, which alleviates some of the constraints that limit the fields achievable by DC magnets. "High field" in the pulsed mode is a function of pulse duration and might be considered as starting around 60 T. (See Box 1.1 for a list of some other magnetic field strengths.)

While this definition of high field will not help the reader decide whether a magnet of one type operating at field *x* is a higher field magnet than a magnet of anther type operating at field *y*, it does make clear why it is difficult to increase the maximum field strength delivered by magnets of any given type. The materials in a high-field magnet of any given type are, by definition, close to failure. One of the objectives of this report is to identify those areas of magnet design and technology where future developments are likely to enable raising the field strengths delivered by high-field magnets, rendering today's high-field magnets "conventional." These developments should make it cheaper to build magnets like the best we have today, promoting their wider distribution.

HIGH-FIELD MAGNETS

As already noted, the construction of high-field magnets has always posed engineering challenges. Solenoids generating fields of about 2 T were built in the 19th century using resistive conductors, and even at fields that low, both the mechanical strength of the materials used and heating were issues. In the 1930s, W.F. Giauque and F. Bitter built water-cooled magnets of novel design from

BOX 1.1 SOME MAGNETIC FIELD STRENGTHS

- In outer space the magnetic flux density is between 10^{-10} T and 10^{-8} T.
- Earth's magnetic field at latitude 50° is 2×10^{-5} T and at the equator (latitude of 0°), 3.1×10^{-5} T.
- The magnetic field of a horseshoe magnet is ~0.1 T.
- The magnets in clinical medical MRI spectrometers operate around 4 T; the current highest field is 9.4 T.
- In a sunspot, the field is ~10 T.
- Strongest continuous magnetic field yet produced in a laboratory:
  —About 20 T with a single superconducting magnet,
  —25 T with a superconducting magnet that has a high-temperature superconducting magnet insert (2003),
  —35 T with a resistive magnet (2002), and
  —45.2 T with a hybrid magnet (2003).
- Strongest (pulsed) magnetic field yet obtained nondestructively in the laboratory: 80 T (1997) for ~10 ms.
- Strongest (pulsed) magnetic field ever achieved (with explosives) in the laboratory (Sarov, Russia), 2,800 T.
- The field on a neutron star is 10^6 T to 10^8 T.
- Maximum theoretical field strength for a neutron star, and therefore for any known phenomenon, is 10^{13} T.

resistive conductors that produced steady-state fields of about 10 T, a big advance. Bitter magnets are still used today. Modern versions produce fields of 30-35 T. While they are relatively inexpensive to build (tens of thousands of dollars), they are costly to operate because of the power they consume and the cooling they require. At fields in this range the energies stored in a magnet of useful size are so large that mechanical failure can have dangerous consequences.

In 1911, H.K. Onnes discovered that many metals become superconducting at temperatures close to 0 K. For magnet designers, superconductivity looked like a godsend. The flow of current through the coils of a superconducting electromagnet generates no heat because there is no resistance, so no cooling is required beyond that needed to maintain its coils in the superconducting state. In addition, once energized, superconducting magnets consume no power and do not have to be connected permanently to a power supply. However, it was soon discovered that no matter how cold they are, the metals in which superconductivity was first demonstrated become resistive when exposed to magnetic fields much lower than those generated by the resistive electromagnets of the day. The quenching of the superconducting state by external magnetic fields occurs in all superconducting materials, not just the metals studied by Onnes. What varies from one superconductor to the next is the field strength at which quenching occurs—i.e., the critical field—and consequently the highest field the material can deliver when it is formed into a magnet.

Only in 1961 were materials discovered that remain superconducting in fields high enough to be interesting to magnet designers, and the use of these materials for magnet fabrication has exploded since then. The number of superconducting magnets operating in instruments in laboratories and hospitals around the world is hard to estimate, but the committee was told by an industry representative that every year manufacturers sell about 2,000 MRI instruments and roughly 500 NMR spectrometers. These instruments contain superconducting magnets collectively worth billions of dollars. Other arenas in which superconducting magnets are used on a large scale are high-energy physics and fusion research. The demand for superconducting wire suitable for high-performance magnets increased enormously in response to the construction of the Large Hadron Collider (LHC) at CERN and will increase even more as construction of the International Thermonuclear Experimental Reactor (ITER) gets under way.

As is explained in the body of this report, the construction of magnets from superconducting wire is a complex art. The performance of all such magnets is limited by the properties of the superconductors from which they are made, especially their critical fields. Mechanical strength and fabricability are also vital issues. These challenges notwithstanding, the maximum strengths of the fields produced by superconducting magnets have gradually increased to about 25 T. Hybrid magnets, which consist of a resistive solenoid inside a superconducting solenoid, can deliver substantially higher DC magnetic fields (about 45 T), but of course they continuously consume power and generate heat in their normal conducting sections.

In 1986, materials were discovered that superconduct at temperatures up to 130 K, much higher than the highest temperature achieved by previously known superconducting materials (about 23 K). These high-temperature superconductors are ceramic copper oxides, which suffer from intrinsically weak links at internal grain boundaries, making the fabrication of magnets from them extremely difficult. They are very interesting to magnet designers, however, because their critical fields are far higher than those of any of the superconductors now routinely used for magnet fabrication. The technical challenges they pose are being overcome, so the field strengths that can be obtained from superconducting magnets are likely to increase significantly in the next few years.

Resistive magnets can generate fields with strengths greater than about 45 T, but only for short times. The NHMFL has magnets that generate fields of about 60 T for tenths of a second, 65 T for hundredths of a second, or about 200 T for milliseconds. If partial or total instrument destruction can be tolerated, fields well above 300 T can be generated for microseconds. As the duration of the field pulse a magnet delivers declines, however, so too does its utility as a tool for scientific research. Consequently, the committee took the view that both the technologies and the science associated with fields of very short duration (less than a few milliseconds) lie outside the scope of its inquiry.[2]

This report has been written for readers who have a technical background and at least some familiarity with high magnetic field science. The committee's decision to write at this level was made following discussions with the NSF. The body of this report begins with an overview of the science that is being done using high-field magnets and the science opportunities and challenges that might open up if higher-field magnets were developed. It closes with a discussion of magnet technology that explains why the fields generated by today's most powerful DC magnets are less than two orders of magnitude stronger than those available to scientists in the 19th century, and points out the opportunities that now exist for developing more powerful magnets. This report includes several appendixes the readers may find useful, such as descriptions of selected high-field facilities around the globe, tutorials on advanced topics, and a glossary of common terms.

[2]Most pulsed magnet techniques offering high fields of duration less than 1 ms are of the destructive variety, where the magnet is designed to explode with each use. For instance, single-turn coil magnet devices driven by a capacitor can generate fields of 100-250 T for 4-8 μs.

2

Scientific Challenges and Opportunities with Higher Fields

Magnetic fields are powerful tools for studying the properties of matter because they couple directly to the electronic charge and magnetic moments of the protons, neutrons, and electrons of which matter is made up. The properties of most materials are only weakly dependent on the strengths of the magnetic fields to which they are exposed, and for these substances, magnetic fields can be used analytically to determine fundamental properties such as their characteristic electronic energy scales and the band structures of metals and insulators, the placement of atoms in molecules, or even the internal structure and dynamics of living creatures. On the other hand, in some materials the magnetic field couples strongly and dramatically influences their properties: for example, in quantum Hall devices, magnetic materials, and superconductors. For these substances, magnetic field strength is as important a thermodynamic parameter as temperature or pressure. Included in this category are many materials important for the production, control, and measurement of high magnetic fields such as high transition temperature (T_c) superconductors. As the committee argues elsewhere, improved understanding of these superconducting materials, which will derive in part from experiments done using state-of-the-art high-field magnets, will lead to the construction of better magnets.

Research using high-field magnets has been remarkably fruitful in the past.[1]

[1]For additional historical context, see National Research Council, *High-Magnetic-Field Research and Facilities*, Washington, D.C., National Academy Press, 1979; National Science Foundation, *Final*

(See Appendix A for a list of Nobel prizes awarded for research that used or significantly affected the development of high magnetic fields.) There is every reason to believe that it will continue to be so, especially if the field strengths of the magnets available to the scientific community continue to increase. In this connection, it is important to note that charged particles move in circular orbits in a magnetic field, the radius of which shrinks as the magnetic field strength increases. Similarly, the smallest size resolved by magnetic moment or spin probes shrinks with increasing field strength. Thus the need to study and characterize ever smaller objects, both those that exist in nature and those fabricated artificially, will not be satisfied unless magnets are fabricated that deliver fields of ever increasing strength and instrumentation is developed that supports their effective use.

Paralleling the distinction made above, this chapter is divided into three sections. It begins with a discussion of high magnetic field research in condensed-matter and materials physics that emphasizes new phenomena that are likely to be revealed and known phenomena that would be better understood if higher fields were available. The chapter continues with a discussion of the impact of high-field magnets on the disciplines of biology, chemistry, biochemistry, and physiology as a result of their use in instruments that exploit nuclear magnetic resonance (NMR). In particular, the committee highlights the impact high magnetic fields have had, and continue to have, on the study of the solution structures of biological macromolecules by NMR, on solid-state NMR of biological and inorganic materials, and on electron paramagnetic resonance (EPR) of metal centers in proteins and catalysts. The committee discusses the impact high magnetic fields have had on two forms of magnetic resonance spectroscopy that have developed since the Richardson report—namely, magnetic resonance imaging (MRI) and ion cyclotron resonance (ICR) mass spectroscopy. In all these areas, magnets that operate at higher fields than those available today would yield large scientific dividends.

CONDENSED-MATTER AND MATERIALS PHYSICS

High-field research in materials science is intrinsically multidisciplinary, merging ideas from physics, chemistry, biology, and engineering, and integrating both theory and experiment. It is pursued predominantly by condensed-matter physicists, the largest subfield within physics today. Materials science, the dominant activity at the world's high magnetic field laboratories, utilizes techniques as diverse as thermal and electrical transport, thermodynamic characterization, magnetization, optical spectroscopy, and magnetic resonance. Many classes of materials are

---

Report of NSF Panel on Large Magnetic Fields, Arlington, Va., National Science Foundation, 1988 (also known as the Richardson report).

investigated, and measurements are done over a wide range of temperatures, pressures, and magnetic fields.

In the 1920s, when high magnetic fields first became available in Europe, they were used initially to investigate simple metals and, later, semiconductors. This work resulted in the first experimental determinations of how individual electrons behave in solids and was extremely influential in the development of the theory of solids between 1930 and 1950. Among the many successes of this synergistic enterprise must be counted the first microscopic explanations of how electrical and thermal transport occur in metals and insulators, and why certain metals become magnetic. This work led ultimately to the development of the science that enabled invention of the first solid-state electronic device, the semiconductor transistor.[2] It would be hard to overstate the impact of these developments on the economies of the industrialized nations in the second half of the 20th century.

Electronic correlations are at the intellectual heart of modern condensed-matter physics. Interactions within populations of electrons lead to emergent collective properties that transcend those of individual electrons, such as superconductivity, magnetic order, and even the formation of the electronic gaps that distinguish metals from insulators. These properties reflect a balance of interactions among the electrons in a population and are strongly affected by differences in dimensionality, crystal symmetry, the spin of constituent atoms, and chemical bonding. Research on correlated-electron systems deals with issues ranging from the most fundamental (e.g., determination of the mechanism responsible for high-transition-temperature superconductivity in copper oxide layered compounds) to the most highly applied (e.g., learning how to control the microstructure of materials so that high-T_c superconductors with the highest possible critical fields can be produced for superconducting magnet construction).

Historically, this field has been constantly refreshed and reinvigorated by the discovery of new materials, such as copper oxide superconductors, heavy fermion magnets and superconductors, organic conductors, and nanoscopic materials such as fullerenes. It has also benefited tremendously from the availability of low-dimensional semiconductor structures of improved quality and purity. As was the case for the noninteracting electron science of the early 1900s, experimental discoveries in this field have had a significant impact on the development of theoretical understanding, which in turn has led to fruitful suggestions about new directions to pursue in materials development. In the past three decades, eight Nobel prizes have been awarded for work in this field (2003, 1998, 1996, 1987, 1985, 1977, 1972,

[2]Indeed, it was a combination of cyclotron resonance and the Hall and de Haas–van Alphen effects that helped characterize the electron transport properties of solids that enabled these inventions.

1970), most recently in 2003 to A. Abrikosov, V. Ginzburg, and A. Leggett for their work on superconductivity and superfluidity. High magnetic field research in advanced semiconductor structures in particular led to the discovery of the integer and fractional quantum Hall effects, resulting in the physics Nobel prizes awarded to K. von Klitzing in 1985 and to R. Laughlin, H. Stormer, and D. Tsui in 1998.

In addition to its intellectual importance, research in correlated-electron systems has already led to numerous technological advances, such as improvements in the sensitivity of the magnetic read heads used for information storage, which depend on the giant magnetoresistance of hybrid magnetic/metallic systems, and the improvements in communication that have resulted from the superior signal-to-noise ratios and interference rejection of high-T_c superconductor filters. The economic promise of this research area is enormous. Improvement in the properties of permanent magnets would impact both the efficiency of electric motors and the density and reliability of magnetic storage media. The quest to understand materials that become superconducting at high temperatures and to discover new materials that superconduct at even higher temperatures has already had important practical results. Improvements in magnetic field sensors and in key electronic components have resulted, as well as the development of high-field inserts for superconducting magnets, which will soon be used for research but may also have bioimaging applications. While currently only at the demonstration stage, superconducting power cables could have a huge economic and environmental impact by reducing power losses in electric transmission networks. Finally, electronic correlations induced by the collapse of metallic screening and finite size effects become increasingly important as the size of electronic components decreases. The trend toward miniaturization has naturally led to an increased interest in nanoscale devices that have novel electronic properties because the devices combine superconducting and magnetic components with more conventional semiconducting components.[3] In every case, progress will be linked to the discovery of materials with improved collective properties.

Understanding how electronic correlations are manifested in the macroscopic behavior of correlated-electron materials is key to the rational design of future generations of advanced materials. This task will require the skills of both experimentalists and theorists from a broad range of disciplines and will need the most advanced tools and techniques. The next section outlines the most important classes of correlated-electron materials and highlights the role high-field measure-

[3]Parallel advances in the speed and miniaturization of electronics have allowed greater exploitation of high fields by enabling experiments in the compact, transient environments offered by pulsed-field magnets.

ments play in developing our understanding of them at both the fundamental and the technological level.

Superconductors, heavy fermion compounds, and organic molecular metals are classes of complex materials in which magnetic, electronic, and structural properties are strongly related. When temperature, pressure, and doping are varied, the existence of multiple phases is often revealed. Paramagnetic, long-range, magnetically ordered, and superconducting phases are seen, which sometimes coexist. In high-T_c superconductors, the reference scale is the transition temperature T_c. In heavy fermion systems, it is the single-impurity Kondo temperature that competes with intersite magnetic couplings, while in organic conductors it is the coupling between chains or planes that often governs other properties.

High-Temperature Superconductivity

The discovery of high-temperature superconductivity in $La_{2-x}Ba_xCuO_4$ ceramics by J. Bednorz and K. Muller in 1986 inaugurated a new era in solid-state physics. Within the next 6 years, the family of high-temperature superconductors had expanded to include Y-, Bi-, Tl-, and Hg-based systems with maximum T_c ranging from 90 to 130 K, respectively, and more than 10,000 scientific papers had been published. Thus in the last decade of the 20th century, high-temperature superconductivity emerged as a major area in physics. Experiments done at high magnetic fields have contributed much to the characterization and elucidation of high-temperature superconductivity and indeed have revealed many of its more remarkable features.

Why is high-temperature superconductivity so important, or, more precisely, why do so many condensed-matter physicists choose to work on this subject? Both fundamental and practical considerations come into play. On the fundamental side, the essence of the challenge is to solve the strong correlation problem. What happens when the electrons in a metal can no longer be described using the noninteracting electron paradigm, L.D. Landau's theory of Fermi liquids? How do electron-electron interactions change a half-filled band, which Landau's Fermi liquid theory indicates should make an excellent metal, into an insulating antiferromagnet?

Nevertheless, the great theoretical challenges presented by the strong correlation problem do not in themselves explain the high level of international activity in this area. There is an additional ingredient—namely, the prospect that we might someday be able to make superconductors that work at room temperature and above. The idea of practical, room-temperature superconductors, with their distinctly quantum mechanical properties, such as the Meissner and Josephson effects, is tantalizing. What we now know about the mechanism of superconductivity in the materials being investigated, which necessarily arises from Coulomb interactions and quantum statistics, suggests that transition temperatures of several hundred kelvin might be possible. It is this combination of fundamental theoretical importance and exciting practical potential that drives the field.

Avenues of Research

All high-temperature superconductors share a key feature that appears to be responsible for their high-temperature superconductivity: the presence of planes containing Cu and O atoms separated by bridging materials that act as charge reservoirs for those planes. These materials become superconducting at temperatures significantly higher than those of the previously known highest-T_c compounds, which are now called low-temperature superconductors: Nb compounds, for instance, have a maximum T_c around 23 K. High-T_c materials also have extraordinarily high upper critical magnetic fields (H_{c2})—for example, 170 T for the widely studied YBCO and maybe 500 T for bismuth- and thallium-based compounds. On the one hand, the high critical fields of these materials make them attractive as conductors for use in high-field magnets, but on the other, their high critical fields are a serious barrier to their full characterization. The critical fields of many of these materials are so high that their normal (nonsuperconducting) states cannot be studied using even the most powerful magnets available today.[4]

Superconductivity in conventional materials is explained by a theory proposed by J. Bardeen, L. Cooper, and R. Schrieffer (BCS theory) and is understood to result from an interaction between electrons and phonons that causes an effective attraction between the electrons, allowing them to pair up. When a conventional superconductor becomes superconducting, the transition to this new, paired state causes a reduction in the potential energy of its charge carriers and a slight increase in their kinetic energy. The net amount of energy released is defined as the condensation energy. In many materials, the electron pairs that result have a fully symmetric internal symmetry, which is a natural consequence of phonon-mediated pairing. In the cuprate high transition temperature superconductors, condensed pairs have a different symmetry, which is indicative of an entirely different pairing mechanism.[5] Thus models for superconductivity in these materials propose a different "glue" for binding carriers together.

[4]That is, studies of the normal state at relatively low temperature; even the HTS materials available today with the highest T_c are not superconducting at room temperature.

[5]The working fluid of superconductors consists of pairs of electrons (or pairs of the holes left behind in a crystal when an electron moves somewhere else). These Cooper pairs form a coherent state with specific symmetry properties.

There are several reasons for the extraordinary focus on high-temperature superconducting (HTS) materials in recent years: their intrinsic scientific interest; the cross-disciplinary nature of the field, which reaches across boundaries that often divide materials scientists and chemists from experimental and theoretical physicists; the potential applications of materials that superconduct at temperatures above the boiling point of liquid nitrogen (77 K); and, finally, the possibility of finding a superconductor that has a critical temperature above room temperature. Applications for HTS materials include filters for cellular phone systems; superconducting transmission lines, generators, motors, transformers, and fault current limiters; higher field MRI instruments and NMR spectrometers; microwave systems; and (of course) magnetically levitated transportation systems.

The scientific challenge posed by HTS materials is more fundamental than simply understanding why they superconduct. The oxide high-T_c superconductors are a family of materials in which even the properties of the normal state are not as well understood as they are for metals like aluminum, lead, or niobium. The identities of the carriers of charge and spin—that is, the HTS equivalents of the electrons and holes in metals, semiconductors, and low-temperature superconductors—are still being debated. Thus one of the key challenges posed by these materials is understanding the physics of their normal states, either at temperatures above T_c or at fields high enough to quench their superconducting states. Since the low-temperature/high-field regime is inaccessible for many of these compounds because current magnets do not deliver fields high enough, most measurements of the HTS normal state have been done above T_c. This approach is often unsatisfactory because thermal energies are so large at those temperatures that the details of the physical phenomena of interest are obscured by thermal fluctuations.[6]

The competing phases of magnetism and superconductivity that exist in HTS materials are illustrated in the phase diagram provided in Figure 2.1. As the carrier doping (number of holes) is increased in these materials (usually by raising the oxygen content), they are transformed from an antiferromagnetic insulator into a metallic superconductor that has a T_c that is also dependent on carrier concentration and peaks at a level termed "optimal doping." While this behavior has been understood for some time now, more recently a pseudo-gap regime has been added to the phase diagram. It is believed that in this range of doping and

[6]Note that transition temperature is not the only driving parameter. One might think that research in the low-temperature/high-field limit might best be done with HTS materials that have low transition temperatures. This is not necessarily so, because these materials are often less amenable to analysis using techniques such as photoemission and optical spectroscopy. Finally, sample purities for the different families of cuprate HTS compounds can vary widely.

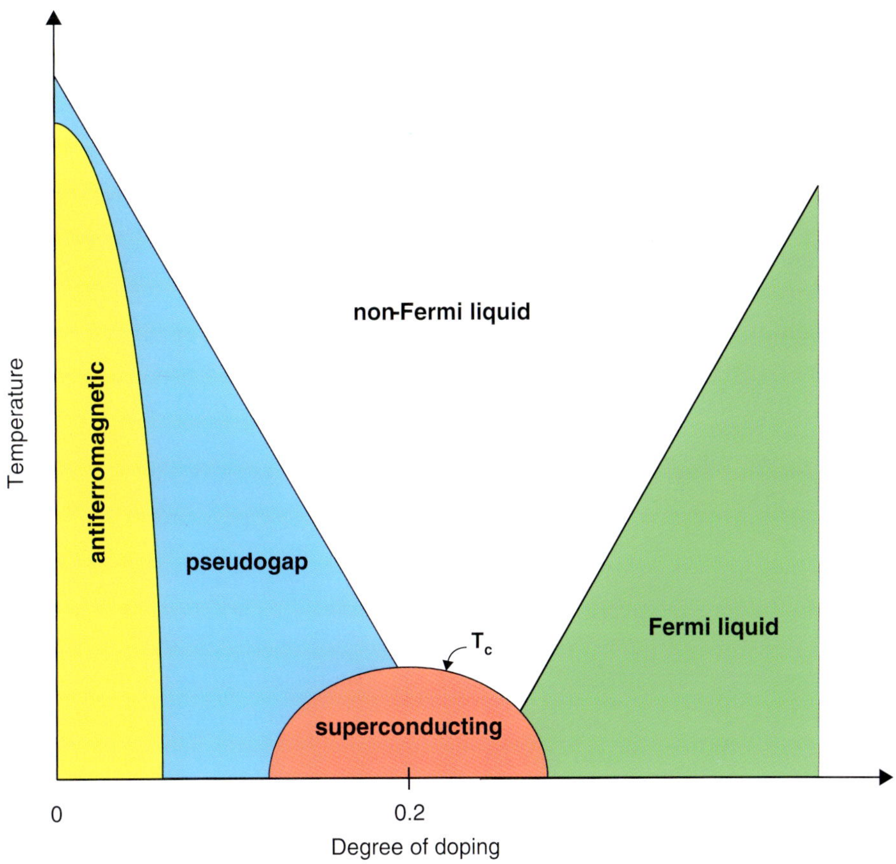

FIGURE 2.1 Generic phase diagram for CuO_2 (cuprate) superconductors, showing temperature versus doping concentration (the latter variable maps onto the order parameter for the material). The properties of the cuprates vary with temperature (*y* axis) and the doping per unit cell of CuO_2 (*x* axis). Theorists are unable to explain why the superconducting transition temperature (thick black line) is so high in the cuprates. However, if they could understand the behavior of the cuprates in the pseudogap region (blue), they might be able to explain high-temperature superconductivity.

temperature (above T_c), the carriers are paired but the pairs do not yet form a superconducting state.

As described in greater detail below, research using high magnetic fields has been critical to uncovering the secrets of high-temperature superconductivity. It is expected that high magnetic fields will continue to be essential as understanding of this phenomenon grows and potential applications are realized.

Although the primary goal of HTS research is to understand the origin of the superconducting state in materials with the highest transition temperatures, much can be learned from the study of HTS materials with lower transition temperatures, because there is every reason to believe that the underlying physics is the same in all of them. The most promising material of this kind is $YBa_2Cu_3O_{7-x}$ ($x \approx 0.5$, $T_c \sim 60$ K). It has approximately the same transition width, and the same level of intrinsic disorder as optimally doped $YBa_2Cu_3O_{7-x}$ ($x \approx 0.05$, $T_c \sim 92$ K). The lower-T_c form has a lower carrier concentration, which should make it an easier material in which to observe quantum oscillations and cyclotron resonance. These phenomena can be used to characterize the shape of the Fermi surface in the normal state and to yield the effective masses of the carriers at the Fermi energy. Experiments with $YBa_2Cu_3O_{6.5}$ in the normal state have additional advantages; the temperature can be stabilized using liquid nitrogen (~77 K).

It is fundamentally important to understand the behavior of the upper critical field, H_{c2}, of HTS materials as a function of temperature, which is why determination of the H_{c2} of HTS materials at low temperatures has become an important issue in HTS research. So far most studies have been limited to fields less than 20 T, which are well below the zero-temperature values of H_{c2} (~100 T). In addition, high magnetic fields may prove invaluable in revealing the nature of the pseudo gap in high-temperature superconductivity, a potential key to understanding the microscopic mechanism of the HTS state. These experiments will require very high magnetic fields—around 100 T—and at lower temperatures, even higher fields may be required.

High magnetic fields (>35-40 T) are a requirement for studying the H-T (magnetic field and temperature) phase diagram of the recently discovered two-gap superconductor MgB_2. (For more discussion of MgB_2, please see the section "Emerging Superconducting Materials" in Chapter 3.) The superconducting energy gap is essentially the energy needed to break the Cooper pairs apart: It also determines the thermodynamic properties of the material and is directly related to the superconducting transition temperature. Most superconductors have just one energy gap, but experiments suggest that magnesium diboride (MgB_2) has two. The gaps correspond to transition temperatures of 15 K and 45 K and combine to give an overall transition temperature of 39 K. A spectacular increase in the upper critical field of MgB_2 was achieved recently by selective alloying of s and p bands with nonmagnetic impurities. $H_{c2}(0)$ values have increased tenfold: from 3 to 5 T for single crystals to 35 T for $H_{\|c}$ and to 50 T for $H_{\|ab}$ for dirty MgB_2 samples. These advances offer an exciting opportunity for studying the novel physics of two-gap superconductivity, nonequilibrium interband phase textures, and vortex dynamics and pinning at high magnetic fields greater than 50 T. Because such studies require sweeps of magnetic field strength, extended averaging times for sensitive measurements, and carefully controlled conditions so that samples can be compared, steady-state fields are generally required.

Opportunities in Vortex Physics

Work on the mechanisms of superconductivity, particularly high-temperature superconductivity, has revealed a rich new scientific landscape. One of the important areas opened up by these activities, vortex physics, is described here in some detail because of its intellectual vitality.

There are two classes of superconducting materials: Type I and Type II. Type I superconductors correspond loosely to the pure element low-temperature superconducting (LTS) materials Hg, Sn, and Pb, in which superconductivity was first discovered. External magnetic fields are fully excluded from the bulk of Type I superconductors by surface currents flowing within the London penetration depth, and the critical fields at which superconductivity is lost (quenched) are very low, less than 0.1 T, which is why Type I superconductors have few commercial applications. Type II materials have much higher critical fields because they enter a mixed state at a lower critical field H_{c1} of ~0.01 T (see Figure 2.2). Above H_{c1}, Abrikosov vortices form, which consist of flux tubes containing a quantum of flux $\phi_0 = 2 \times 10^{-15}$ Wb. Each field filament is encircled by a supercurrent flowing to a depth equal to the London penetration depth. The centers of these vortices can be viewed as normal cores where the superconducting order parameter is suppressed. When a current is driven through the material, the flux lines experience a Lorentz force that tends to push them perpendicular to the current. If they so move, the process is dissipative and introduces resistance. This phenomenon is well understood, but exactly how magnetic fields penetrate into superconductors, how the flux lines move, and how they interact with defects in the material are not well known. The flux lines can also repel each other, so flux flow is a complex, many-body effect. We are now coming to understand that flux flow is like other dissipative effects such as earthquakes and avalanches, and the physics of vortex motion has many parallels to other areas of physics.

This quantized flux-tube state was first conceived of by A. Abrikosov to describe the situation that occurs when the superconducting coherence length ξ is much shorter than the penetration depth λ. He found that the vortex state was both quantized and characterized by a lattice structure, the existence of which was subsequently confirmed by neutron diffraction, magnetic decoration, and magneto-optical and transmission electron microscopy experiments. A particular curiosity of Type II superconductors is that the interface energy between their normal and superconducting regions is negative, making fine-scale subdivision of the vortex state energetically favorable. The vortex density depends on magnetic field as $(\phi_0/B)^{0.5}$. Bulk superconductivity is destroyed when the normal cores of the flux tubes in a material overlap, which occurs at a field H_{c2} of $\phi_0/2\pi\xi^2$. H_{c2} values can be remarkably high. For LTS materials such as Nb-Ti or Nb_3Sn, the values are

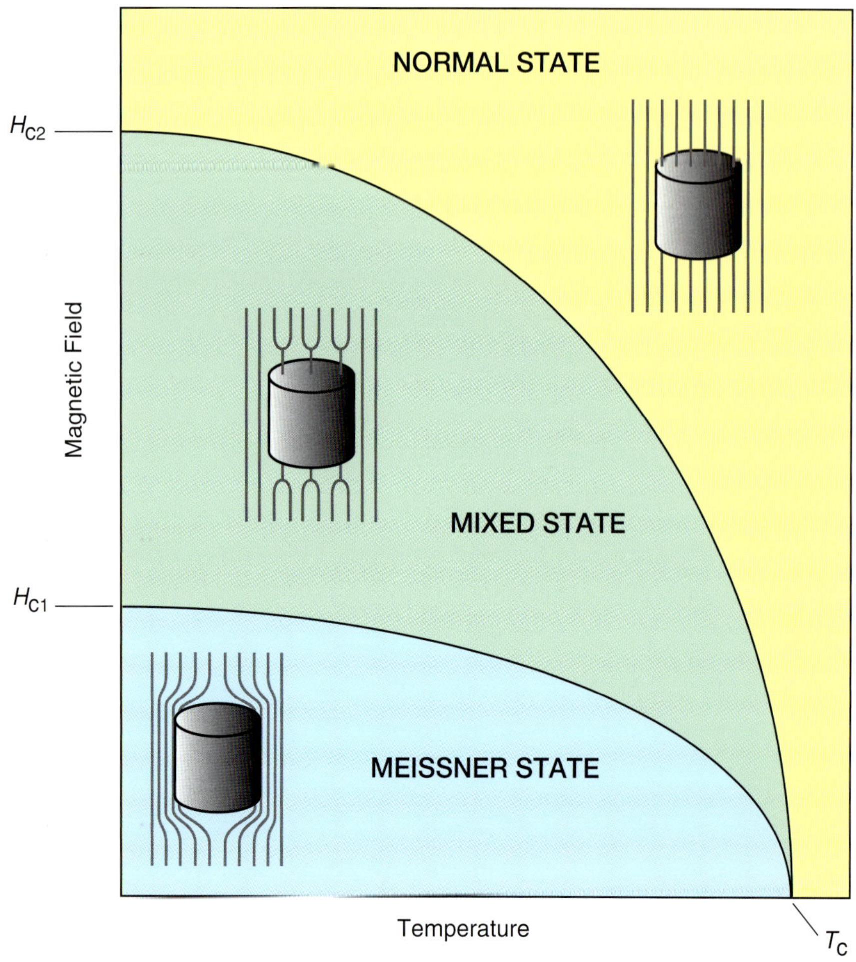

FIGURE 2.2 Phase diagram for penetration of the magnetic field into a Type II superconductor. The green mixed-state region is where Abrikosov vortices are formed and vortex physics comes into play.

about 15 T and 30 T, respectively, but for cuprate superconductors with high T_c, H_{c2} can exceed 100 T.

LTS and HTS Type II superconductors differ in the degree of interaction among the vortices with each other and with defects in the material; HTS materials offer a much richer spectrum of physics. In the essentially isotropic LTS (Type II) metallic superconductors, vortices are line objects with significant line tension and are thus effectively pinned by even dilute microstructural defect arrays, provided that their size is optimum for interacting either with the vortex core size of diameter 2ξ or the screening currents that flow on a scale of the penetration depth λ. Since $\xi << \lambda$ for all useful Type II superconductors, the engineering of pinning centers became an exercise in nanotechnology years before the word "nanotechnology" passed into common use. In optimized Nb-Ti wires, about 20 percent of the volume of ~3 nm-thick nonsuperconducting ribbons of α-Ti are dispersed in a superconducting matrix of Nb-Ti solid solution. These normally conducting ribbons pin vortices very effectively, allowing transport currents to develop that are about 10 percent of the depairing current density that would destroy superconductivity altogether. The strong vortex pinning in such wires allows highly stable currents to flow with virtually no dissipation, making them ideal for the fabrication of superconducting magnets.

By contrast, the highly anisotropic nature of the layered cuprate HTS materials makes the line tension of their vortices weak. Indeed in many of the cuprates, vortices break up into pancakes that lie on the cuprate planes, which are strongly superconducting. Correlations along the vortex line tend to be very weak, which makes thermal activation easy and allows a wide range of excitations that are essentially absent in the metallic superconductors. For instance, vortices can be set in motion by the Lorentz force from an externally applied transport current, enabling studies of driven phases, steady-state motion, and the new area of dynamic phase transitions.

In traditional superconductors, thermal energy is limited to about 20 K by the superconducting transition temperature, and the vortex tubes form an elastic solid. High-temperature superconductors offer a new possibility: Thermal energies up to about 100 K may melt the vortex solid, creating a novel liquid state with dramatically different properties that arise from the relative motion of vortices. As early as 1988, motion of vortices below T_c was found to create undesired dissipation in high-T_c cuprates. Theorists soon realized that vortex phases and phase transitions embody many fundamental features of condensed-matter physics, including reduced dimensionality, entanglement of flexible line objects, and the role of disorder in elastic media.

The nature of the vortex physics therefore depends not only on static variables such as pinning energy (controlled by defects in the material) and coupling energies (controlled by the anisotropy of the material) but also on dynamical variables, including temperature, magnetic field, and applied current. For instance, the density of vortices in a superconductor depends on the applied magnetic field; for an average HTS material, the separation between vortices is about 80 nm at 1 T, 6 nm at 50 T, and about 4 nm at 100 T. Adjustments to magnetic field and temperature can scan a broad range of potential interactions. The HTS materials therefore exhibit a wide range of vortex interactions that give rise to many new phenomena.

Vortices control many aspects of the electromagnetic response of HTS materials. The richness of their vortex phase diagrams is caused by the fact that four energy scales (tilt, shear, pinning, and thermal energies) are of the same order of magnitude in the field regime determined by the properties of HTS materials. Virtually all studies so far have been made in weak fields of up to about 10 T, but much of the most interesting vortex physics, for instance, occurs in concentrated vortex arrays in YBCO at 25 to 50 T.

The nontrivial and nonlinear pinning of vortex dynamics—vortex creep—and vortex lattice melting at an irreversibility field H_{irr} much smaller than H_{c2} has determined many aspects of the direction of HTS research. Even though many features of the macroscopic electromagnetic behavior of HTS materials have been successfully explained using vortex concepts, new theoretical predictions (including states of vortex matter such as vortex liquids, quasi-ordered lattices, and Bose glasses) continue to open avenues for experimental research. Understanding and controlling vortex pinning is particularly important for applications since strong pinning is required to develop the high critical current densities needed for driven- and, especially, persistent-mode NMR and MRI magnets.

These research avenues are addressed in greater detail in Appendix H.

Outlook for High-Temperature Superconductivity

It follows that the challenge presented by HTS materials today is both scientific and technological. The scientific challenge persists, despite the vast improvements in measurement and instrumentation capabilities that have been made since, say, the 1960s, when LTS materials were explored. For example, the BCS electron-phonon coupling can be measured quantitatively in tunneling spectroscopy experiments carried out between a polycrystalline film of Al and one of Pb. In contrast, the Bi HTS surface has been studied using scanning tunneling microscopes so stable that the same atoms can be measured over periods of weeks or months, with tunneling spectroscopy carried out atom by atom. Nevertheless, despite the power of such modern experimental techniques, HTS materials give up their secrets only very slowly.

For those interested in understanding HTS materials, it is an unhappy fact that over the past 50 years, field strengths of the magnets available for experimental work have not kept pace with the increase in the critical fields of the superconductors being discovered. The critical fields of Nb-Ti and Nb_3Sn were measured using Bitter magnets in the 1960s, but the critical fields of many of the HTS materials being investigated today are not accessible with the magnets available today. One approach to the limitations imposed by this mismatch has been to study versions of the materials of interest, e.g., copper oxides, that have lower T_c

and H_{c2}. YBCO, which is widely studied, can be modified to have more accessible T_c and H_{c2} by reducing its oxygen content. For obvious reasons, such work-arounds are only partly satisfactory; it is the materials with the highest T_c and highest H_{c2} that offer the ultimate challenge and the ultimate opportunity. It is paradoxical that the development of the HTS materials that may make it possible to someday manufacture magnets delivering unprecedented high fields is being hindered because only magnets built of such HTS materials are likely to deliver fields high enough for their characterization.

Neutron experiments have played a crucial role in HTS work by providing insights into momentum- and energy-dependent spin gaps, as well as an increasingly precise overview of the spin fluctuations, which if not at the root of the superconductivity, are responsible for many of the anomalous normal state properties of HTS cuprates. Recently, a series of experiments done in Europe provided a remarkably clear demonstration of the value of linking the best available superconducting magnet technology with modern neutron instrumentation. These experiments revealed the antiferromagnetism of the vortex state in underdoped cuprates, as well as the sensitivity of the spin fluctuations in optimally doped samples to the vortex melting line, which had previously been seen only using bulk measurements sensitive to very long range phenomena. Additional work showed that a prominent singlet-triplet excitation (the resonance peak) for YBCO is also very sensitive to the presence of vortices. NMR experiments provided important early hints about the *d*-wave nature of high-T_c superconductors as well as their pronounced antiferromagnetic fluctuations. Most recently, work following the neutron experiments revealed the antiferromagnetic nature of the vortices in LSCO, suggesting similar behavior in YBCO.

Higher magnetic fields would make it possible to perform experiments beyond the upper critical fields of some of the higher T_c cuprates. These experiments would lead to an improved understanding of at least the phenomenology of materials that are more useful technologically. In addition, it would be tremendously valuable to enhance the capabilities available for doing various spectroscopies and microscopies at higher fields. Much has been learned from neutron and scanning tunneling microscopy at relatively modest fields in the last decade, and the big challenges of the next decade will be observing how vortices merge to form either a strange metal or an insulator using either position-resolved tunneling or momentum-resolved magnetic neutron scattering, or both. Other interesting opportunities lie in microwave and optical spectroscopies. Experiments of this kind would not just produce information similar to that already available, they would also make it possible to visualize terra incognita, the quantum phase transition believed to separate the high-temperature superconductor from its true zero-temperature parent phase. The committee points out that beyond providing a new window on a still unsolved problem, these capabilities would greatly facilitate experimentation with other materials of high current interest: most quantum magnets, which might be hosts for protected qubits, as well as narrow bandgap semiconductors of potential use in magnetoelectronics, applications that take advantage of not only the electron's charge but also its spin.

Ultimately, while the specific compounds discussed above provide good examples of exciting new materials that need to be investigated using high magnetic fields, it must be emphasized that they are not the only ones. All high-temperature superconductivity is ripe for high-field experimentation, and the higher the fields, the better. Generally speaking, pulsed fields with lengths on the order of 10 ms are required to provide sufficient measurement time for many studies; many other analyses will require the precision and stability obtainable only using steady-state fields. The opportunity to unlock the science of high-temperature superconductivity using high-field magnets is matched only by the opportunity to exploit this knowledge to design HTS materials that can meet the demands of the industrial marketplace.

Heavy Fermion Systems

The ability to understand and, ultimately, to predict the circumstances under which materials become magnetic is important at both a fundamental and an applied level. Nowhere is this more starkly evident than in the class of metallic magnets called Kondo lattices, or heavy fermion systems. Heavy fermions are compounds containing rare earth elements such as Ce or Yb or actinide elements such as U. Their (inner shell) conduction electrons often have effective masses (known as quasi-particle masses) several hundred times as great as that of free electrons, resulting in low Fermi energies. This property makes them reluctant superconductors. Yet at cryogenic temperatures, many of these materials are magnetically ordered, others show strong paramagnetic behavior, and some display superconductivity through mechanisms that transcend traditional BCS theory. Research suggests that Cooper-pairing in the heavy fermion systems arises from the magnetic interactions of the electron spins rather than from lattice vibrations. In this section the committee briefly describes the key topics in this quickly growing area of research and comments on the role that high magnetic fields can play.

All heavy fermion systems contain an ordered lattice of localized magnetic moments, generally due to the presence of rare earth or actinide elements, but systems based on transition metals are also studied. These moments interact with itinerant electrons contributed by other constituents of these intermetallic compounds. Much of the interest in these materials is spurred by the belief that their collective properties are well described by a generic phase diagram that is intriguingly similar to the generic phase diagram that characterizes high-T_c materials,

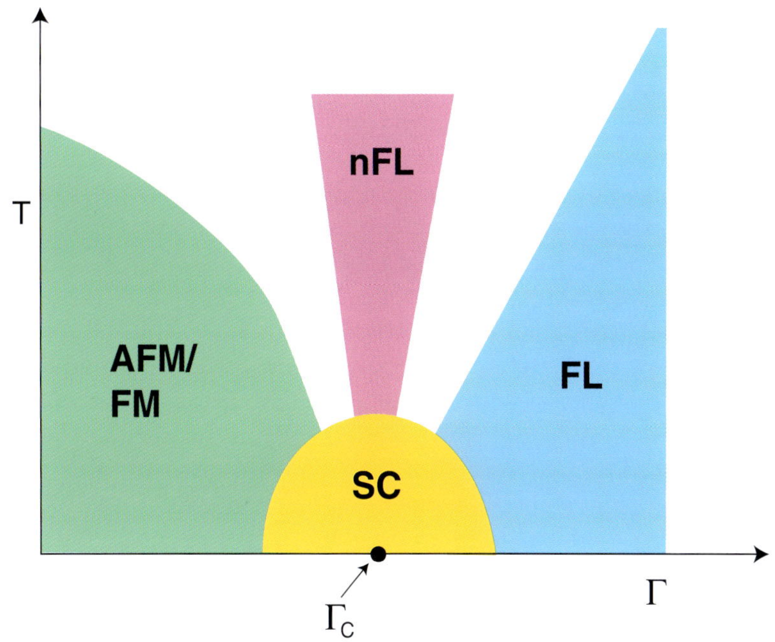

FIGURE 2.3 Generic phase diagram for heavy fermion systems in the plane of temperature (*T*) versus the order parameter (Γ). The superconducting phase (SC) surrounds the quantum critical point (Γ_C); on the right is the normal-state Fermi liquid phase (FL), with the pink region describing the non-Fermi liquid regime. Finally, the left-hand green area maps out the ferromagnetic and antiferromagnetic regimes.

despite the inherently different natures of these two classes of materials (see Figure 2.3).[7] The phase diagram describes the stability of magnetic order as a function of an external variable G (also called the order parameter) such as pressure, chemical composition, or magnetic field. The dominant feature of the phase diagram is its quantum critical point at $\Gamma = \Gamma_C$ and 0 K, where magnetic order gives way to a mass-enhanced, nonmagnetic, metallic state from which this class of materials gets its name: heavy Fermi liquids, or simply heavy fermions. The quantum critical point in this phase diagram has attracted interest because some materials arc superconductive when they are in this part of the phase diagram. The proximity of this superconducting state to the magnetically ordered phase suggests

[7]The interested reader may find more detail in two excellent reviews: G.R. Stewart, Non-Fermi liquid behavior in d- and f-electron metals, *Rev. Mod. Phys.* 73, 797 (2001) and A.C. Hewson and D. Edwards, *The Kondo Problem to Heavy Fermions,* London, England: Cambridge University Press, 1997.

that the superconductivity observed is unconventional, mediated perhaps by magnetic rather than lattice fluctuations. Further, a new type of universal, but still unconventional, "non-Fermi liquid" behavior can be found in the vicinity of the quantum critical point that is associated with anomalous critical fluctuations.

Measurements done at high magnetic field have contributed substantially to our understanding of these systems by providing a means for studying the underlying electronic structures and by serving as a tuning parameter to adjust the underlying energy scales and, finally, as a tool for forcing the material into states that cannot be accessed at zero field. Since the experimental record is vast, the committee focuses here on a few illustrative examples.

The forces responsible for the suppression of magnetic order, and the generation of the quantum critical point in the phase diagram of Figure 2.3, remain a topic of debate. It is of central importance to determine whether the electrons responsible for the localized magnetic character found at small values of Γ leave the local moment sites and become itinerant as Γ approaches Γ_C.

The remarkably robust superconductivity of the metal alloy (and heavy fermion system) $CeCu_2Si_2$ has been a long-standing mystery to condensed-matter physicists. One of the many triumphs of the BCS theory of superconductivity was its simple explanation for the strong suppression of the superconducting transition temperature T_c by magnetic impurities in metals. The local moments act with opposite sign on the two electrons of a spin singlet Cooper pair and hence are pair breaking. A by-product of this idea is the expectation that metals such as heavy fermions, whose low-temperature properties are strongly renormalized by magnetic effects, would not become superconducting. It was a shock, then, when superconductivity was discovered in $CeCu_2Si_2$ in 1979 by F. Steglich and colleagues in Dresden, Germany, only shortly after the existence of heavy fermion materials themselves was recognized.[8] The discovery ignited research into the physics of these heavy fermion metals.

Magnetization experiments performed on $CeRu_2Si_2$ determined that a magnetic field of 7.8 T would drive this material from a nonmagnetic metallic state ($\Gamma > \Gamma_C$) into a magnetically ordered state ($\Gamma < \Gamma_C$).[9] Quantum oscillations have been observed in the magnetization as the field varies, establishing the presence of itinerant electrons at all fields.[10] However, there is a sudden change in the frequency

[8] F. Steglich, J. Aarts, C.D. Bredl, W. Lieke, D. Meschede, W. Franz, and H. Schafer, Superconductivity in the presence of strong Pauli paramagnetism: $CeCu_2Si_2$, *Phys. Rev. Lett.* 43, 1892-1896 (1979).

[9] F.S. Tautz, S.R. Julian, G.J. McMullan, and G.G. Lonzarich, The nature of elementary excitations below and above the metamagnetic transition in $CeRu_2Si_2$, *Physics* B 206-207, 29-32 (1995).

[10] H. Aoki and S. Uji, Transition of f-electron nature from itinerant to localized: Metamagnetic transition in $CeRu_2Si_2$ studied via the de Haas-van Alphen effect, *Phys. Rev. Lett.* 71, 2110 (1993) and Tautz et al., 1995 (see above).

of the oscillations as the field exceeds 7.8 T, indicating that previously itinerant electrons have become localized on Ce moment sites at high field. While the 7.8 T is a field easily achieved by superconducting magnets in the laboratory, the success of such quantum oscillation studies depends critically on the availability of a wide range of magnetic fields for experimental use, both above and below the critical field. Subsequent experiments in which compositional variation was used to vary Γ through Γ_C suggest that the Fermi surface restructuring seen is a general feature of quantum critical points and is responsible for the stabilization of magnetic order in a number of metallic magnets.

Ongoing research seeks to establish whether this result can be extended to different sorts of magnetic phase, and this information can only be obtained from quantum oscillation studies, particularly if the quantum critical point must be generated using external magnetic fields. Higher fields will be required for such quantum oscillation studies as more highly itinerant systems based on transition metals begin to be investigated. For instance, the unconventional magnetic superconductor UPt_3 undergoes a similar moment delocalization transition at 20 T, and the same transition occurs in URu_2Si_2 near 40 T. Studies of these systems will have to await the availability of more powerful magnets equipped to study samples at dilution refrigerator temperatures.

As the experiments on $CeRu_2Si_2$ imply, magnetic fields can also be used to tune the stability of magnetism in intermetallic compounds, where electronic correlations are extremely strong and emergent energy scales are very small. Recent measurements of the $YbRh_2Si_2$ system provide perhaps the most dramatic demonstration of this tuning effect.[11] In zero field, $YbRh_2Si_2$ is antiferromagnetically ordered below 0.065 K. Application of a 0.05-T field suppresses the magnetic order to 0 K and produces a nonordered state with anomalous properties. Its specific heat diverges logarithmically with temperature. Its resistance is linear at temperatures between 10 mK and 10 K, and its magnetic susceptibility is weakly divergent. At higher fields, normal nonmagnetic and metallic behavior is regained in $YbRh_2Si_2$. These observations suggest that materials tuned to the vicinity of a quantum critical point have special properties, perhaps originating with the critical fluctuations of the incipient magnetic phase. Indeed, it has been possible to show that resistance and magnetization are jointly controlled by the ratio of the field to the temperature, confirming this hypothesis. The committee notes in passing that

[11] P. Gegenwart, J. Custers, C. Geibel, K. Neumaier, T. Tayama, K. Tenya, O. Trovarelli, and F. Steglich, Magnetic-field induced quantum critical point in $YbRh_2Si_2$, *Phys. Rev. Lett.* 89, 056402 (2002) and J. Custers, P. Gegenwart, H. Wilhelm, K. Neumaier, Y. Tokiwa, O. Trovaerlli, C. Geibel, F. Steglich, C. Pepin, and P. Coleman, The break-up of heavy fermion electrons at a quantum critical point, *Nature* 424 (6948), 524-527 (2003).

if a material is found that is naturally tuned so that in zero field and at low or zero pressure it is at its quantum critical point, its linear magnetoresistance could be very attractive for sensor applications. Although the field required to tune $YbRh_2Si_2$ to the quantum critical point is modest, similar observations have been made with different Kondo systems at fields as large as 35 T. As the field scale is material-specific, it is virtually inevitable that new quantum critical systems will be discovered as higher fields become available and as a wider range of high-field measurements makes more detailed analysis of high-field states possible.

In order to understand quantum criticality it is imperative that we discover and examine different types of magnets. These investigations will require an arsenal of instruments that operate at ever higher fields and ever lower temperatures. The instruments available are being pushed to the limit today, and to meet future needs, improved magnets, both DC and pulsed, and improved ancillary instrumentation will be required. Pulsed fields can provide access to different parts of the phase diagram, but transient effects (such as sample heating due to the magnetic field) and the need for precision measurements still drive the need for steady-state measurements in DC fields. The most detailed work requires excellent and simultaneous control of both temperature and field—very difficult to achieve with pulsed fields. However, such demanding conditions are often not required to establish properties of phase diagrams, to determine global features in insulating compounds, or (most notably) to get Fermi surface information via quantum oscillation measurements.[12] A complementary approach using both pulsed and steady-state fields is necessary to identify new phenomena and then to study them in appropriate detail.

High magnetic fields are a particularly powerful means for suppressing collective instabilities that make it difficult to study the underlying disordered state from which the ordered state emerges. This approach is particularly illuminating when applied to investigations of the superconducting ground state, especially if the electrons in the superconducting condensate have unconventional pairing, but is also crucial for understanding the response of superconductors to high fields, which is essential if practical use is to be made of them. The way magnetic fields suppress unconventional superconductivity is unusually interesting. One example of a novel, field-driven state is the Fulde-Ferrell-Larkin-Ovchinnikov (FFLO) state. In Type II superconductors, the field–temperature phase diagram shows a steady penetration of vortices into the superconducting material, leading to the termination of the superconducting state at a characteristic critical field. In superconductors of

[12]During the brief lifetime of the 60-T long-pulse magnet at NHMFL in Los Alamos it was convincingly demonstrated that experimental measurements of properties such as heat capacity and magnetostriction are very possible in pulsed fields of this sort.

high purity and large electronic mean free paths, at sufficiently high fields, the superconductor may accommodate the increasing Zeeman energy of the Cooper pairs through a spatial modulation of the superconducting order parameter, involving layers of superconducting material separated by magnetic domain walls. Two different groups have recently reported signs of a transition from a conventional Type II superconductor to this novel FFLO superconducting state, as well as transitions in applied field between states of differing orbital angular momentum in the layered Kondo superconductor $CeCoIn_5$ at temperatures around 0.8 K and fields over 10 T.[13] Although there is mounting evidence that the superconductors in the Kondo magnets are unconventional, their superconducting transition temperatures are very low, limiting their technological importance. A more comprehensive description of superconductors in high fields appears above in the section on high-T_c superconductors.

High fields can also be used to destabilize other types of collective states in metallic magnets. For instance, an insulating ground state has been observed in zero field transport and inelastic neutron-scattering measurements on the Kondo lattice system $Ce_3Bi_4Pt_3$. It was unknown whether the insulating gap implied by these measurements was a simple band structure effect, or whether it was the result of interactions among electrons quasi-localized near the Ce moments. In a first experiment of its kind, heat capacity measurements carried out using the Los Alamos 60-T long-pulse magnet showed that magnetic field suppressed the insulating gap, yielding a high-temperature state that was a normal metal.[14] This experiment unambiguously demonstrated that the Kondo gap state is collective, resulting from magnetic interactions between itinerant electrons and localized moments.

Finally, there are certain collective states in metallic magnets that occur only in high magnetic fields. One striking recent example is provided by the heavy fermion system URu_2Si_2, where in zero field there is an unusual orbital ordering transition at 17 K, followed by a superconducting transition at lower temperature. When high magnetic fields are applied it is found that at 36 T, the Zeeman splitting of the conduction electron states suppresses the orbital order, and a reentrant magnetic state is stabilized between 36 and 39 T. This material is unique among metallic

[13] A. Bianchi, R. Movshovich, C. Capan, P. G. Pagliuso, and J. L. Sarrao, Possible Fulde-Ferrell-Larkin-Ovchinnikov superconducting state in $CeCoIn_5$, *Phys. Rev. Lett.* 91, 187004 (2003), and H.A. Radovan, N.A. Fortune, T.P. Murphy, S.T. Hannahs, E.C. Palm, S.W. Tozer, and D. Hall, Magnetic enhancement of superconductivity from electron spin domains, *Nature* 425 (6953), 51-55 (2003).

[14] M. Jaime, R. Movshovich, G.R. Stewart, W.P. Beyermann, M.G. Berisso, M.F. Hundley, P.C. Canfield, and J.L. Sarrao, Closing the spin gap in the Kondo insulator $Ce_3Bi_4Pt_3$ at high magnetic fields, *Nature* 405 (6783), 160-163 (2000).

systems in that a relationship between orbital order and superconductivity may be inferred, and unique also in that this relationship apparently protects the system against the establishment of itinerant magnetic order, a common instability for this class of materials. The most direct assessment of orbital and magnetic order can be obtained from x ray and neutron scattering measurements, and they have been used to great effect in other classes of materials, most notably transition metal oxides, where the instabilities are present in zero field.

Research on heavy-fermion and related systems constitutes a major part of the National High Magnetic Field Laboratory (NHMFL) research portfolio. Researchers from around the world come to use the magnets, but it is less often that U.S. researchers go overseas to conduct this sort of research. The leading exception is for critical scattering experiments with high fields. The high-field capabilities for x-ray and neutron scattering are vastly superior in Europe, so U.S. researchers must go there for these sorts of experiments, at a considerable investment of their time and resources. These sorts of experiments are impossible in the United States today for systems such as URu_2Si_2, where magnetic and orbital instabilities are driven by high fields and can then be examined with beams from scattering sources.

Low-Dimensional Semiconductors

High magnetic fields have traditionally played an important role in the study of electronic properties of semiconductors. Even today, decades after the first breakthroughs, high magnetic fields play a crucial role in investigations of the electrical and optical properties of heterostructures, quantum dots, and superlattices. The semiconductor physics community now sees a clear need for magnetic fields well in excess of 30 T, combined with low temperatures, to explore the new physics associated with new materials and structures processed on the nanoscale. The availability of these high fields will also open up new fundamental areas for study—for example, systems where electron-electron interactions dominate—and will be of great assistance in the characterization, further development, and exploitation of new semiconductor materials. In this section, the committee focuses on compelling opportunities in the strongly correlated electron systems offered by low-dimensional semiconductors.

Research in low-dimensional electron systems in semiconductors is driven by more than just scientific curiosity. Many of the structures being investigated could become the building blocks for new solid-state device technologies. They might be used, for example, to produce useful devices based on novel electronic phenomena, such as Bloch oscillations in modulated structures or Coulomb blockades in quantum dot systems. They are also prime candidates for use in the currently exciting fields of spintronics and quantum computing. Studies of low-dimensional semiconductor systems at high magnetic fields provide information regarding their charge transport, energy level structure, and spin states and the way electronic interactions engender these properties. Such information is crucial for the successful use of such systems in any potential application.

In the last 15 years, there have been many exciting developments in the science of low-dimensional semiconductors. In low-dimensional semiconductors charge carriers are confined to the interface between two semiconductors that have different bandgaps. They are often modulation-doped, which is to say that they have dopant atoms that are placed in the higher bandgap material at some distance from the interface, while their mobile charge carriers (electrons or holes) reside in the smaller bandgap material in a plane very near the interface. Since their charge carriers can essentially move freely in that plane, the system is effectively two-dimensional (2D). The spatial separation between the 2D carrier plane and the dopants (ionized impurities) significantly reduces the scattering of carriers by dopant atoms, making such systems ideal for studying electron-electron interactions in two dimensions. This effect is particularly interesting at low temperatures and high magnetic fields, where the kinetic energy of the carriers is quenched and Coulombic interactions between carriers dominate the properties of the system.

Investigation of the properties of 2D semiconductors at low temperatures and high magnetic fields is one of the richest areas in condensed-matter physics today. Such studies have already led to the discovery of the integral and fractional quantum Hall effects (IQHE and FQHE) and to two Nobel prizes in physics (1985 and 1998). As elaborated below, electrons in such "flatlands" continue to reveal unanticipated phases and new phenomena, which derive from subtle, nonintuitive electron-electron interactions (see Figure 2.4). By confining the carriers in the lateral directions, various one-dimensional (quantum wire) and zero-dimensional (quantum dot) carrier systems can also be formed, which are equally interesting.[15]

Over the years, studies of 2D carrier systems in semiconductors at high magnetic fields have led to the discovery of new states of matter. The FQHE is a good example. The carrier system involved is an intrinsically many-body, incompressible quantum liquid that, in the limit of zero temperature, can flow without dissipation. It is described by the Laughlin wave function, a many-electron function that has the Coulomb repulsion between the electrons built in, keeping them far apart. The electrons described by the FQHE wave function exhibit short-range correlation

[15]For additional background on these topics, please see S.D. Darma and A. Pinczuk, eds., *Perspectives in Quantum Hall Effects: Novel Quantum Liquids in Low-Dimensional Semiconductor Structures,* John Wiley and Sons, New York, 1997, or S. Girvin, The quantum Hall effect: Novel excitations and broken symmetries, *Topological Aspects of Low Dimensional Systems,* A. Comtet, T. Jolicoeur, S. Ouvry, and F. David, eds., Springer-Verlag, Berlin, 2000.

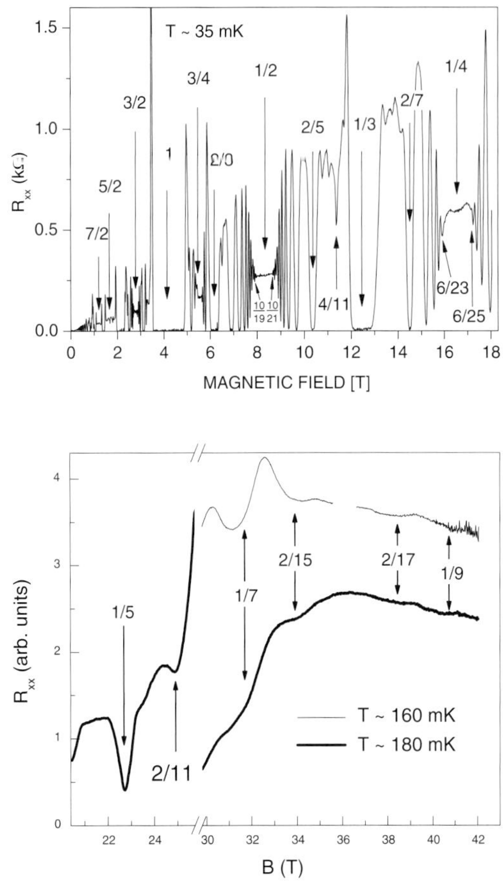

FIGURE 2.4 Magnetotransport data for a state-of-the-art 2D electron system in a GaAs/AlGaAs heterostructure. The FQHE is a fascinating manifestation of collective behavior in a 2D system of strongly interacting electrons. At particular magnetic fields, the electron gas condenses into a remarkable quantum liquid state, which flows without dissipation in the limit of zero temperature: It has vanishing resistance and a quantized Hall voltage. This state is very delicate, requiring high-quality material with a low carrier concentration and extremely low temperatures. The key feature is the strong systematic dependence of the resistance on the applied magnetic field; the detailed structure is attributable to different collective phases of the electrons. *Above:* Data taken in a superconducting magnet showing a wealth of features, many of which provide new insights and twists in the many-body physics of 2D electrons. *Below:* Data on the same sample at higher fields and higher temperatures, providing information on the transition from a Wigner solid state to a liquid of fractional quantum Hall states. Together, these plots show how magnetic fields can be used to control and explore new phenomena in low-dimensional semiconductor systems. [See W. Pan, H.L. Stormer, D.C. Tsui, L.N. Pfeiffer, K.W. Baldwin, and K.W. West, Transition from an electron solid to the sequence of fractional quantum Hall states at very low Landau level filling factor, *Phys. Rev. Lett.* 88, 176802 (2002) for details.] Figures courtesy of W. Pan, Sandia National Laboratory.

in their positions (hence they behave as a liquid), but they do not form a crystalline solid phase and do not possess long-range order. Until its experimental discovery in 1982, this phase of matter was entirely unknown.

When the *Final Report of NSF Panel on Large Magnetic Fields* appeared in 1988, only two states of 2D electrons at high field were known: the IQHE and FQHE states. The report emphasized the need for higher magnetic fields to find out if other 2D phases exist. A particular example cited was the Wigner crystal state, a state in which the electrons minimize their Coulombic repulsion by forming an ordered array. This state was expected to occur in extremely high-quality (low-disorder) samples at very low Landau level fillings ($\nu \lesssim 1/5$) and very high magnetic fields ($\gtrsim$ 30 T). Since 1987, remarkable progress has been made in the production of semiconductor crystals by molecular beam epitaxy at Bell Laboratories (Lucent Technologies) and elsewhere. Among other things, this progress has resulted in the production of GaAs/AlGaAs-based heterostructures of superb quality with very high mobilities. The study of these materials in instruments that combine higher DC magnetic fields and lower temperatures than were previously available has led to the discovery of a remarkable number of new electron states and phenomena (see Figure 2.4 (bottom) for examples of the magnetoresistance data that can be obtained from a state-of-the-art GaAs/AlGaAs 2D electron system). These phenomena arise from the collective interactions of the electrons in 2D systems and include the following:

- Composite fermions,
- Skyrmions,
- Even-denominator FQHE states,
- Ferromagnetic phases and transitions between QHE states, both in the integer and the fractional regimes,
- Striped phases at large half-integer fillings,
- Reentrant QHE and insulating phases at high fillings,
- Insulating states at very low fillings, suggestive of a pinned Wigner crystal phase,
- Radiation-induced zero-resistance states at low fields, and
- Various phenomena in bilayer systems, including bilayer QHE states stabilized by interlayer interaction and phase coherence.

These tremendous discoveries, some completely unexpected, have far exceeded the expectations of most experts. Each has added a new twist to the rich physics of interacting electrons in a high magnetic field. For example, a new electron state with a highly anisotropic in-plane conductivity was discovered very recently at half-integer Landau level fillings (e.g., at $\nu = 9/2$). This is believed to be an interaction-induced "stripe" phase, composed of narrow, parallel regions of QHE states with $\nu = 4$ and $\nu = 5$. Another example is the emergence of skyrmions near the $\nu = 1$ filling factor. These are low-energy excitations of the ferromagnetic QHE ground state at $\nu = 1$ and are charged, finite-size, electron spin textures. Again, the Coulomb interaction is responsible for bringing about these excitations.

These discoveries have depended on the availability of high-quality samples and high magnetic fields. Equally crucial has been the availability of electronically and mechanically quiet environments for making observations. Progress in this field will continue in parallel with improvements in sample quality. It is clear, however, that as the quality of samples improves, lower temperatures and better instrumentation will be needed to explore their ever more subtle properties. Ample access to state-of-the-art magnets will also be important, because the experiments necessary to reveal new phenomena will be time-consuming explorations of the properties of these materials as a function of many variables—for example, sample and carrier density, temperature, and sample orientation with respect to the direction of magnetic field.

Much can be learned from the study of one-dimensional (1D) and zero-dimensional (0D) systems at high magnetic fields and low temperatures. Experiments of this kind will be especially revealing when the magnetic length becomes comparable to the system size. Since the length scale of a magnetic field is inversely proportional to the square root of its field strength, fields on the order of 100 T are needed to probe scales of about 2.5 nm. These measurements can provide information about the interplay between electron-electron interactions, confinement, and magnetic energies in small systems containing only a few electrons. Electronically quiet environments will be even more important for this research than they are for 2D systems, because 1D and 0D systems are unusually sensitive to mechanical and electronic noise.

The very high fields offered by pulsed magnets are helpful in uncovering entirely new phenomena in these materials, but longer-term magnetic field stability is required for detailed studies of the underlying mechanisms. In summary, the advance of research in 2D electron systems as well as 0D and 1D systems hinges importantly on (1) substantially improved access to the fields presently available, (2) improved environmental stability (including lower electronic noise levels), (3) greater access to lower temperatures, and (4) availability of magnetic fields beyond 30 or 40 T.

Organic Conductors and Superconductors

Organic superconductors are interesting experimental systems because they are well understood chemically and because their low Fermi energies and lower dimensionality favor correlated electron ground states and novel types of superconductivity. Experimentally, using equipment available today, one can observe their properties at magnetic fields well above those necessary to quench their superconducting—or other correlated electron—states and thereby explore quantum oscillations and probe their Fermi surfaces. In this way, the fundamental properties of the charge carriers can be explored and current theories of correlated electrons can be tested.

Charge transfer salts composed of organic molecules are among the most interesting electronic materials ever discovered.[16] Many of their unusual properties are only evident at high fields. Their molecular architecture leads directly to lower electron densities, smaller bandwidths, and larger electronic anisotropies than those found in their inorganic cousins. Nevertheless, their strong Coulomb interactions make them strongly correlated electron systems, and they exhibit virtually all of the cooperative ground states associated with such systems: superconductivity, magnetism, non-Fermi liquid behavior, as well as additional instabilities usually found in low-dimensional systems—for example, QHE, spin and charge density waves, and localization from disorder and from interactions. Their electrochemical growth yields clean materials with very large mean free paths (1-10 mm) and hence strong orbital effects in magnetic field.[17]

Early experiments on the quasi-1D TMTSF (tetramethyltetraselenafulvalene) salts showed that in open-orbit electronic systems, the magnetic field anisotropically shrinks the electron wavefunction, resulting in a field-induced dimensional crossover. This is manifest as a metal-antiferromagnetic insulator transition, or field-induced spin density wave. The anisotropy of these materials made it possible to separate the orbital and spin effects associated with their superconductivity and pointed to a triplet spin pairing state.

The discovery of a class of quasi-2D charge transfer salts (known as bis(ethylenedithio)tetrathiafulvalene, or BEDT-TTF, salts) with closed orbits has led to a flurry of high magnetic field studies. In both quasi-1D and quasi-2D materials, long mean free paths lead to qualitatively new magnetotransport

[16]For an introduction to these topics, see J. Singleton and C. Mielke, Quasi-two-dimensional organic superconductors: A review, *Contemp. Phys.* 43, 63-96 (2002), or S. Chakravarty and P.W. Anderson, Interlayer Josephson tunneling and breakdown of Fermi liquid theory, *Phys. Rev. Lett.* 72, 3859 (1994).

[17]S. Uji, T. Terashima, H. Aoki, J.S. Brooks, M. Tokumoto, N. Kinoshita, T. Kinoshita, Y. Tanaka, and H. Anzai, Fermi-surface reconstruction in the organic conductor $(BEDT\text{-}TTF)_2(TlHg(SCN)_4)$, *J. Phys. Condens. Matter* 6, L539-L547 (1994); O. Klein, K. Holczer, G. Gruner, J.J. Chang, and F. Wudl, Electrodynamics of the superconducting state of kappa $(BEDT\text{-}TTF)_2Cu(NCS)_2$, *Phys. Rev. Lett.* 66, 655-658 (1991).

phenomena. In particular there are enormous angular-dependent magnetoresistance oscillations, which allow direct measurements of the Fermi surface shape and parameters. A wealth of Fermi surface instabilities and reconstructions allow quantitative investigation of nesting vectors, spin effects, and coupling parameters. There are, additionally, giant de Haas van Alphen and Shubnikov-de Haas oscillations, which can be fit in sufficient detail so that electron-phonon interactions can be characterized and the nature of the superconducting state explored. With these clean, low-electron-density, anisotropic samples, it has even been possible to observe the quantum Hall effect in bulk crystals.

Recent discoveries are as exciting as those that preceded them. For example, field-induced superconductivity has been observed in a BETS salt (bis(ethelynedithio)-tetraselenafulvalene), a most beautiful example of the Jaccarino-Peter effect, i.e., compensation of an externally applied magnetic field by an internal exchange field. There are also several hints that the inhomogeneous FFLO superconducting state exists in BEDT salts. There are even experimental indications of the existence of a new state of a field-induced charge density wave, analogous to the FFLO state, where spin splitting competes with gapping of the Fermi surface.

These experiments have typically pushed the capabilities of magnet facilities to their limits of high field, low temperature, high pressure, and exact alignment of the field with the anisotropic crystal axes.

Combining High Fields with X-Ray and Neutron Scattering

Neutron scattering has been at the forefront of research into magnetic materials since 1948, when the first direct observation of antiferromagnetism was made by C.G. Shull and J.S. Smart.[18] Beginning in the 1960s, the observation of collective magnetic excitations by neutron scattering provided direct information on exchange interactions, soft magnon modes and phase transitions, magnon-phonon interactions, and other phenomena. Collective magnetic excitations were observed in many different materials, including metals, which were itinerant ferromagnets where the existence of such excitations had been questioned. More recently, neutron-scattering experiments at high magnetic fields have provided information about quantum magnetic systems, especially at low dimensionality, leading to new insights into many theoretically predicted phenomena, such as the collective effects leading to a gap in the excitation spectrum of 1D spin-1 chains (the Haldane gap). Work in Europe at the highest fields available has shown that high magnetic

[18]C.G. Shull and J.S. Smart, Detection of antiferromagnetism by neutron scattering, *Phys. Rev.* 76, 1256 (1949).

fields induce striped antiferromagnetic order in some high-T_c materials, a key observation for the theory of these materials.[19]

While the use of x-rays for magnetic scattering studies is of more recent vintage, the rapidly increasing brightness of synchrotron sources has made many new kinds of experiments possible. In 1988, D. Gibbs et al. showed the resonance and polarization behavior of x-ray scattering by studying the spiral magnetic structure in Ho.[20] Resonant magnetic x-ray scattering was later used by M.B. Salamon to investigate the magnetic properties of the induced moment of Lu in a Dy-Lu alloy, where the atomic selectivity afforded by the energy dependence of the resonance allowed direct observation of the Lu scattering in the presence of the much stronger scattering from Dy. High photon fluxes also have allowed studies of surface magnetism and the change in magnetic behavior as a function of depth below the surface.

When higher magnetic fields become available at synchrotron light sources, many new scientific opportunities will open up. For example, the hole-doped cuprates have high critical fields, and the instruments available today do not allow observations to be made on these materials at interesting ratios of applied to critical field. In the area of heavy fermions, where the range of fields currently available is not adequate for exploring the quantum critical points of many systems, instruments operating between 20 and 25 T would open important new possibilities. The same range of fields would make it possible to study induced magnetic moments in nonmagnetic materials, both molecular and metallic (such as Lu). In this case, there are exciting opportunities for x-ray studies because the atomic specificity provided by such resonant scattering should make it possible to isolate the magnetic signal component—for example, to observe the Lu contribution in a Lu-Dy alloy. Furthermore, there are 3D quantum systems where the interesting physics begins at 20 T, a field that is not available at an x-ray- or neutron-scattering facility anywhere in the world. An equally important opportunity would be the pursuit of structural and dynamic properties in complex organic and biological systems. High fields, when coupled with neutron and x-ray sources, will provide a new variable-contrast agent for pursuing these studies. The possibility of polarizing unpaired electrons is similarly exciting.

[19]B. Lake, H.M. Rønnow, N.B. Christensen, G. Aeppli, K. Lefmann, D.F. McMorrow, P. Vorderwisch, P. Smeibidl, N. Mangkorntongstar, T. Sasagawastar, M. Noharastar, H. Takagistar, and T. E. Mason, Antiferromagnetic order induced by an applied magnetic field in a high-temperature superconductor, *Nature* 415, 299-302 (2002).

[20]D. Gibbs, D.R. Harshman, E.D. Isaacs, D.B. McWhan, D. Mills, and C. Vettier, Polarization and resonance properties of magnetic x-ray scattering in holmium, *Phys. Rev. Lett.* 61, 1241-1244 (1988).

Current Capabilities

During the past two decades, major user facilities for x-ray and neutron scattering were constructed in the United States and older user facilities were upgraded. The Advanced Photon Source at Argonne National Laboratory and the Advanced Light Source at Lawrence Berkeley National Laboratory (LBNL) are world-class, third-generation synchrotron radiation sources equipped with a wide range of instrumentation. The Stanford Synchrotron Radiation Laboratory at the Stanford Linear Accelerator Center, the National Synchrotron Light Source at Brookhaven National Laboratory, and the Cornell High Energy Synchrotron Source have also been steadily upgraded. Instrumentation at the Center for Neutron Research of the National Institute of Standards and Technology has improved significantly since 1989, instrumentation at the High Flux Isotope Reactor at Oak Ridge National Laboratory is being upgraded now, the source and the instrumentation at the Lujan Scattering Center at the Los Alamos Neutron Scattering Science Center in Los Alamos National Laboratory have been greatly enhanced, and the Intense Pulsed Neutron Source at Argonne National Laboratory continues to operate reliably with a first-class suite of instruments. The Spallation Neutron Source (SNS), now under construction at Oak Ridge National Laboratory, will be the world's most intense neutron source when it begins operation, and it will eventually be equipped with a comprehensive suite of instruments. These developments will place the United States at the forefront of scattering research.

Some provision has been made at U.S. scattering facilities for users who want to study the effect of magnetic fields on their samples, but as Table B.2 in Appendix B documents, the United States has yet to take advantage of the research opportunities that would be created by combining the world-leading capabilities of the NHMFL in magnet design and construction with the outstanding x-ray and neutron capabilities it already has in place. In user surveys done at U.S. scattering facilities, the top concern of the materials science and condensed-matter physics community is sample environment and, in particular, the availability of magnetic fields. Current capabilities limit the range of properties that can be studied at these photon- and neutron-scattering facilities and are especially confining in the field of high-T_c systems, where researchers cannot conduct studies at fields near the critical field strength.

At both the European Synchrotron Research Facility in Grenoble (x rays) and the Hahn-Meitner Institute in Berlin (neutrons), the facilities for doing scattering/diffraction research at high magnetic fields—both those now in place and those planned—are far superior to any in the United States. For example, the neutron-scattering center at the Hahn-Meitner Institute, which already has capabilities unmatched at any U.S. facility, has submitted a proposal for a 40-T magnet[21] and

[21] Installation of the 40-T magnet has been delayed for funding reasons.

is proceeding with plans for a 25-T superconducting magnet, and the European neutron-scattering community has identified the 40-T magnet as its highest priority. Similarly, the FELIX user facility in the Netherlands already combines a high-power subpicosecond laser with 16-T fields. (See Table B.2 in Appendix B for a survey of selected efforts around the world to incorporate high-field instruments with premier scattering centers.) Using these capabilities, European researchers have recently made important advances in probing the mechanisms of high-temperature superconductivity. For instance, neutron experiments in the presence of magnetic fields have revealed the antiferromagnetism of the vortex state in underdoped cuprates.

Seizing the High-Field Scattering Opportunity

The United States would have a world-leading capability for studying the responses of materials to high magnetic fields with neutrons if it developed a suite of instruments with magnetic field capabilities at SNS. Ideally this set of instruments would enable users to do single-crystal diffraction, powder diffraction, small-angle neutron scattering, engineering (residual strain and preferred orientation) studies, neutron reflectometry, and inelastic scattering, all at high magnetic fields. By carefully planning the locations of such instruments, it should be possible to keep stray magnetic fields from interfering with other instruments at SNS. In considering needs at SNS, one should not forget the opportunities that exist for improving high magnetic field instrumentation at existing neutron facilities. Capabilities comparable to or exceeding those now available in Europe could be achieved at those facilities also. Both at SNS and at existing facilities, many of the needs of the user community could be met with off-the-shelf 20-T superconducting magnets, and more aggressive goals might be satisfied with resistive DC magnets operating in the 25-T range. In any case, these developments will require close interagency cooperation and planning, which will have to begin immediately if high-field instrumentation is to be available at SNS when it becomes operational. The White House Office of Science and Technology Policy interagency working group on neutron-scattering facilities is one entity organizing and prioritizing these efforts.

U.S. photon sources are beginning to provide magnetic field capabilities, and requests for such capabilities are beginning to turn up. The Advanced Photon Source at Argonne is planning a workshop where the opportunities provided by high magnetic fields will be explored. Although, unlike neutrons, x-ray photons do not couple directly with the magnetic properties of the materials with which they interact, their disadvantage in this regard is compensated for by the extraordinary brightness of the photon beams now available at synchrotron light sources. Even though the study of magnetic systems with x rays is a relatively young enterprise, it is clear that new opportunities exist.

A significant investment would have to be made to make state-of-the-art magnetic-field experimentation possible at U.S. x-ray and neutron user facilities, but as has already been made clear, the payoff would be more than commensurate. Not only would the capital costs for constructing the equipment required be considerable, it would require a long-term commitment of new operating funds. The committee heard estimates that it might take \$1 million or \$2 million per year to staff a state-of-the-art magnet facility for 24-7 operation at a neutron or x-ray facility, while electric power costs for a steady-state magnet at the highest fields would be on the order of \$100,000 per day, assuming the need for a 30-40 MW power supply and corresponding cooling facilities. While a pulsed magnet would be cheaper to build and operate, the mismatch between neutron pulsing rates (20-200 Hz) or x-ray source pulsing rates (MHz) and magnet pulsing rates (1 Hz) would result in severe performance penalties. The trade-offs will have to be studied in detail before they can be properly evaluated. Furthermore, the costs of making high magnetic fields available at neutron and x-ray facilities are probably so high that the number of such systems that can be built will be limited, and care will have to be taken to ensure they are installed where they can be used most effectively. It is important to note that although the number of high-field magnet suites that can be built at U.S. scattering facilities is certain to be limited by economic considerations, there is a strong need for building at least a few of them. The kinds of measurements they would enable cannot easily be made by scientists who can make only occasional, short-duration visits to facilities in other nations; such measurements require much time. Additionally, these high magnetic field tools would leverage investments that have already been made at existing U.S. resources and would greatly expand domestic science capabilities.

The committee notes that interest has been expressed in solving this problem by buildng a new light source, either a synchrotron or a free electron laser, at NHMFL. In its estimation, this approach is unlikely to be a wise one. First, the nation should invest in high-field instrumentation at its neutron sources before it turns to high-field instrumentation at synchrotron sources. Second, it is likely to be much cheaper to bring high-field instrumentation to existing synchrotron and neutron sources than to build an entirely new synchrotron light source at NHMFL, both in terms of construction cost and long-term operating costs.

Future Needs

The phenomena described above illustrate the amazing range of behaviors displayed by correlated electron systems that are either driven by, or can be probed by, high magnetic fields. This research area is active and vital, and future advances will come from the availability of higher fields and a broader range of precision high-field measurement techniques.

The scale of characteristic fields in correlated-electron systems differs considerably from material to material. So far, most experimental work has concentrated on systems with low characteristic field scales and has involved experiments that use slowly varying magnetic fields. Studies of these systems would benefit greatly from higher field superconducting magnets, an expanded hybrid magnet program, and also the development of high-field, long-pulse magnets. However, equal attention must be paid to improving both the variety of techniques that can be employed at high field and the availability of these techniques to members of the materials science community. Today, we are very far from being able to study any high-field instability in the same detail as an instability that occurs in zero field, largely because of the difficulties that surround the execution of high-field experiments over a wide range of temperatures or in complex environments, such as those supplied by high-pressure cells. Particularly pressing is the need for the development of quantitative magnetometry; improved techniques for carrying out thermal conductivity and heat capacity measurements in pulsed field magnets; better methods for making low-noise, low-signal-magnitude electrical transport measurements, especially at low temperatures; and the integration of general-use pressure and uniaxial stress cells into high-field environments. Furthermore, investment in equipment for producing low temperatures in magnets of all types—superconducting, Bitter, and pulsed—would have an enabling effect on this endeavor, where strong correlations lead to intrinsically small energy scales and the current focus on quantum mechanical effects increasingly directs attention toward the lowest temperatures and the largest values of the field strength:temperature ratio. Finally, it should be noted that high magnetic fields share a strong connection with nanoscience and technology: High fields are both enablers of and enabled by developments at the nanoscale. Various magnetic lengths, such as the flux lattice constant, decrease with increasing magnetic field, allowing true nanometer-scale probes. Similarly, detailed behaviors that can be controlled at the nanoscale could be used in the generation of substantially higher field magnets.

As already noted, nowhere is the gap between the instrumentation available for experimentation at zero field and that available for high-field experimentation wider than in neutron and x-ray scattering. This is particularly unfortunate because scattering experiments provide a powerful means for elucidating atomic and magnetic structure as well as determining the nature of the spatial and dynamical correlations in materials. The development of new high-field capabilities at x-ray- and neutron-scattering centers in the United States could have an enormous impact.

High-Field Facilities for Materials Research

Because the scientific and technological importance of magnetism is widely appreciated, there are centers focused on high magnetic field research throughout the world. The committee identified 32 such centers, including those in the United States, all of which offer access to outside users to some degree. Table B.1 in Appendix B catalogs those centers and summarizes the instrumentation available at each; Table B.2 in Appendix B outlines some of the magnetic field capabilities at selected scattering centers. Appendix B also includes short descriptions of the capabilities at these centers and their research interests.

Just by virtue of the number of its magnet stations, the NHMFL is by far the most important facility in the United States offering users access to high magnetic fields (see Appendix B). The NHMFL has led the United States to world leadership in the production and use of high magnetic fields in many areas of materials research, but it would be unwise to assume that its success is a sure sign of U.S. dominance in all other areas of high-field research. In the first place, materials science is only one of many scientific disciplines that use high magnetic fields. In the second, centers outside the United States, both those that provide services to outside users and those that do not, give stiff competition across the entire spectrum of high-field research.

MAGNETIC AND ION CYCLOTRON RESONANCE: APPLICATIONS OF HIGH FIELDS TO BIOLOGY, CHEMISTRY, AND MATERIALS RESEARCH

Over the past 50 years, analytical techniques based on magnetism have become essential tools for chemists, biomedical scientists, and physicians. While these techniques have contributed to progress in solid-state physics and materials science, their development has also given rise to a community interested in high-field magnets that is not only distinct from the solid-state and materials science community but also far larger, measured either by the numbers of scientists involved or by the resources invested. On the whole, unlike the materials science community, the analytical community does not patronize user facilities. Its members instead do their science using instrumentation available locally.

Many of the magnet-dependent analytical techniques of interest here exploit NMR. NMR is used by chemists and biochemists to determine molecular structures. Solid-state physicists and materials scientists use it to characterize materials. Physicians use an NMR-based technique called MRI to visualize the interior of the human body. An entirely different magnetic phenomenon, ICR, is the basis of a powerful technique for measuring atomic and molecular masses. The current maximum field strengths for magnets in resonance applications vary depending on the physical size of the magnet, which in turn is dictated by the kinds of measurements for which it is to be used. These magnets must be extremely stable and homogeneous. For NMR, field limits are now about 21 T. For ICR, field strengths are just passing 12 T; for MRI, fields are from 4 to 9 T.

The magnet at the heart of the modern NMR spectrometer, MRI instrument, or ICR mass spectrometer is a superconducting magnet, often one having modest field strength but highly optimized with respect to stability and homogeneity. However, the performance of almost all such instruments improves with field strength so that the demands of the large communities that use them have been and continue to be an important stimulus for the development of superconducting magnets of ever-increasing field strength.

Introduction to NMR

Because the nuclei of many atoms—for example, ^{1}H, ^{13}C, ^{15}N, and ^{17}O—have magnetic moments, they absorb and emit radiation at characteristic radio frequencies when placed in magnetic fields. This phenomenon, NMR, was first detected in atomic beams at Columbia University in 1938, and NMR signals were first observed in condensed matter at Harvard and Stanford in 1946. Initially, NMR was used by physicists to study the interactions of nuclei with applied magnetic fields, with one another, and with their environment. Its subsequent use as a tool for investigating the electronic and magnetic properties of materials led to many advances in solid-state physics—for example, the first experimental verification of the BCS theory of superconductivity, by L.C. Hebel and C.P. Slichter in 1959.[22] This classic experiment measured nuclear spin lattice relaxation times in normal and superconducting aluminum. The difference in the temperature dependence of nuclear relaxation and ultrasonic absorption measurements confirmed a central feature of BCS theory, namely, that electrons of opposite spin and momentum are correlated.

The frequencies of the NMR signals produced by atomic nuclei are sensitive to the chemical environment, and in the 1950s, chemists discovered that molecular structure and dynamics could be investigated fruitfully by NMR spectroscopy. It is now routinely applied to virtually everything chemists study, such as inorganic compounds, organic compounds, synthetic polymers, natural products, and fossil fuels. NMR spectrometers are essential components of the modern chemical laboratory.

[22] L.C. Hebel and C.P. Slichter, Nuclear spin relaxation in normal and superconducting aluminum, *Phys. Rev. Lett.* 113, 1504-1519 (1959).

In the late 1950s, biochemists began applying NMR to biological macromolecules like proteins and nucleic acids. The number of magnetically active nuclei in the average biological macromolecule is large, and the complex NMR spectra of these molecules could not be resolved using 1950s instrumentation, so progress was slow. Beginning in the mid-1970s, however, three important developments transformed this field. First, superconducting magnets suitable for NMR were developed that had field strengths greater than 9 T. NMR spectrometers built around such magnets had both the spectral resolution and the sensitivity required for analyzing the spectra of small macromolecules. Improved performance resulted not only from the superior resolution and sensitivity afforded by increased field strength but also from improvements in detector electronics. Second, multidimensional NMR techniques were invented that made it possible to spread the NMR signals produced by macromolecules over several dimensions, which further improved resolution. At the same time, efficient methods were developed for determining which atoms in a molecule are close together and for identifying those that are neighbors because they are chemically bonded. Third, molecular biologists and biochemists devised methods for preparing biological macromolecules in which ^{12}C and ^{14}N are replaced by their spin-1/2 isotopomers ^{13}C and ^{15}N and/or in which ^{1}H is replaced by ^{2}H, either to simplify proton spectra or to sharpen the resonances produced by other nuclei. The result of these developments is that the NMR spectra of macromolecules containing thousands of magnetically active atoms can now be assigned and their structures solved. The present-day molecular weight limit for complete protein structure determination is 40-50 kDa. A recent example is shown in Figure 2.5.

Along the way an entirely different use was found for NMR. If a suspension of cells, a piece of tissue, or even an intact organism is placed in an NMR spectrometer, the spectrum observed will be dominated by the resonances of the low molecular weight metabolites it contains. These metabolites tend to be present at high concentrations, and their resonances are much narrower than those from the macromolecular materials present. Since NMR is a noninvasive technique, in a properly designed experiment, which may involve isotopically labeled nutrients, metabolic events can be followed in a living organism in real time, and often processes occurring in one organ of an animal can be distinguished from those occurring in another.

Since powerful new NMR techniques continue to be developed at a rapid rate and instrumental capabilities continue to improve, it is impossible to predict the limits of NMR spectroscopy. Its scientific impact can be gauged from the number of NMR-related Nobel prizes awarded over the years. The prize in physics was given to I.I. Rabi in 1944 and to F. Bloch and E.M. Purcell in 1952, both for the initial demonstrations of NMR. In 1991, R.R. Ernst received the Nobel prize in

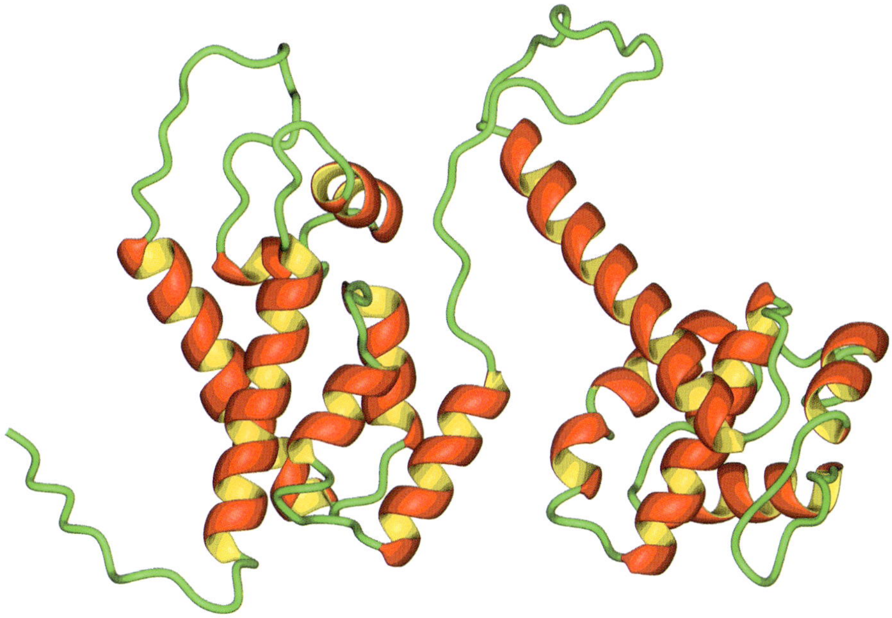

FIGURE 2.5 Ribbon representation of the molecular structure of a 283-amino-acid portion of the Gag polyprotein encoded by the HIV-1 genome, determined by multidimensional solution NMR methods. This structure illustrates both the complexity and the biomedical relevance of problems addressable by modern NMR. This portion of HIV-1 Gag contains two helical domains that are ultimately processed to form the capsid and matrix proteins of the mature HIV-1 virus. [See C. Tang, Y. Ndassa, and M.F. Summers, Structure of the N-terminal 283-residue fragment of the immature HIV-1 Gag polyprotein, *Nat. Struct. Biol.* 9, 537 (2002) for details.] Image courtesy of R. Tycko, National Institutes of Health and M. Summers, University of Maryland at Baltimore County.

chemistry for developments in Fourier-transform NMR and two-dimensional NMR, and in 2002, his colleague K. Wüthrich received the Nobel prize in chemistry for protein structure determination by NMR. In 2003, the Nobel prize in physiology and medicine was awarded to P. Lauterbur and P. Mansfield for the invention of MRI.

Today, NMR is one of the two primary means of determining complete structures of molecules (biological and otherwise), the other being x-ray crystallography. Information about the structures of biological macromolecules is essential for understanding biological processes and for the development of new drugs and new technological materials. In addition, NMR is an important tool for characterizing molecular motions in atomic-level detail, on timescales from $<10^{-9}$ s to >1 s.

Molecular motions are an essential part of every biological process, and they determine the mechanical properties of materials. MRI is one of the most widely used medical imaging modalities (along with x-ray and ultrasound imaging and positron emission tomography (PET) and computer-assisted tomography (CAT) scans), and functional MRI (fMRI) is rapidly becoming the principal method for investigating both normal neurological function and neurological diseases.

NMR is typically performed on molecules either dissolved in a liquid, known as solution NMR, or in solidified samples—for example, crystallized proteins—known as solid-state NMR. The key difference between the two is that molecules in solution are freely diffusing and those in the solid state are not. In solution studies, the extraction of structural information can be hampered by dynamic processes that cause the shape of a molecule to vary on NMR timescales. However, the tumbling motions of molecules in solution average out certain systematic effects that can make it difficult to interpret the results of solid-state experiments.

The Role of Field Strength in NMR

In a typical NMR measurement, a sample that has come to equilibrium in a magnetic field is subjected to a series of radio-frequency pulses that perturb the orientations of its nuclear spins, and the radio-frequency radiation the sample emits in response is measured. The frequency of the NMR signals observed, which is proportional to the energy of the transitions they represent, depends linearly on magnetic field strength. The proportionality constant that relates frequency to field, which is called the nuclear gyromagnetic ratio, is different for each isotope. For hydrogen, the NMR frequency in megahertz (MHz) is ~42.6 times the field strength in tesla (T), but for ^{13}C, the NMR frequency is ~10.7 times the field strength. Thus, the NMR signals produced by different isotopes are easily distinguished by frequency. State-of-the-art NMR spectrometers suitable for macromolecular applications operate at magnetic fields up to 21.1 T (900 MHz for hydrogen) and have bore diameters in the 54-89 mm range.

The magnet in a modern NMR spectrometer is invariably a superconducting magnet constructed using Nb-Ti and/or Nb_3Sn wire, and it operates at 4.2 K or below.[23] The magnets in NMR spectrometers designed for the analysis of liquid samples must produce fields that are much more stable and homogeneous than the magnets used for any other kind of experiment. Typical magnetic field drift rates are less than 10 parts per billion per hour, and fields are homogeneous to less

[23]High-T_c materials have not yet been used in commercial NMR magnets but are likely to be components of NMR magnets built in the future that operate at fields significantly above 21.1 T.

than 10 parts per billion over volumes of about 1 cm^3. These requirements derive from the fact that the NMR signals of individual nuclei within molecules in solution can have intrinsic frequency widths (linewidths) on the order of 0.1 to 10 Hz. Thus, even small magnetic field instabilities and inhomogeneities can broaden the signals significantly, a smearing that becomes unacceptable when the molecule being examined contains many nuclei of the same isotope, each producing a signal that differs in frequency only slightly from that of its neighbors.

NMR spectrometers that operate at high fields are generally superior to those that operate at low fields. Higher-field instruments resolve complex spectra better and have better signal-to-noise ratios. In the NMR spectra of molecules in solution, the signals from chemically distinct atoms of a given isotope have slightly different frequencies due to differences in the magnetic shielding provided by the electron clouds in which they are embedded. These site-specific variations in NMR frequencies are called chemical shifts, and if they did not exist, there would be little reason for chemists to use NMR. Chemical shift differences (in frequency units) increase in proportion with magnetic field, so the bigger the field, the better resolved the spectra obtained, provided that line widths do not increase significantly with field, which is often the case for samples in solution. In an N-dimensional NMR spectrum, the number of resolvable sites—and hence the molecular weight and complexity of the molecule that can be analyzed—in principle should increase as H_0^N, where H_0 is the field strength of the spectrometer's magnet.

The smallest number of atoms or molecules in a sample that can be detected by an NMR spectrometer—its sensitivity—is also a crucial issue in nearly all NMR measurements. Because the energies of the photons an NMR spectrometer must detect are very small, ~0.004 meV or less, ordinary samples must include more than 10^{16} molecules if they are to give NMR signals strong enough to allow detailed analysis. For many potentially interesting systems—for example, integral membrane proteins that function as hormone receptors—the required sample quantities greatly exceed what can be readily prepared. Novel techniques for sensitivity enhancement in NMR are under experimental investigation to address this problem.

In practice, signal-to-noise gains as a function of applied field strength have typically been closer to linear, implying only quadratic reductions in measurement times, largely because the efficiency of the circuitry available for signal detection tends to fall as frequency increases.

The field dependences of nuclear spin interactions can lead to less obvious but equally significant advantages for spectrometers that operate at higher fields. For example, the transverse relaxation optimized spectroscopy (TROSY) effect leads to anomalously large improvements in both sensitivity and resolution of certain classes of experiments important for protein structure determination as fields approach 20 T. Also, when quadrupolar nuclei—nuclei with intrinsic spin angular momentum quantum numbers greater than 1/2—are studied by NMR in solids, the inverse dependence of quadrupolar line broadening on H_0 and the linear dependence of chemical shifts on H_0 produce quadratic enhancements in spectral resolution and sensitivity.

Finally, at high fields, the assumption that sample behavior is independent of field strength begins to break down even for solutions of diamagnetic molecules like proteins. The magnetic susceptibility of most biological macromolecules is sufficiently anisotropic so that their tendency to orient in solution in external magnetic fields is detectable in high-field NMR spectrometers. When molecules orient this way, through-space nuclear magnetic dipole-dipole interactions, which at lower fields are averaged to zero by molecular tumbling, become large enough to measure. Valuable information about molecular structure can be obtained from the analysis of residual dipolar couplings.

Recent Developments in Solution NMR

Although the development of spectrometers incorporating magnets of ever-increasing field strength has been the leitmotif of NMR technology since its inception, other advances, such as those already mentioned in electronics and computation, have also contributed to progress. Another advance is exemplified by the recent introduction of cryogenically cooled probes, which are NMR-signal detection devices in which the radio-frequency coil and preamplifier are cooled to very low temperatures to reduce the contribution of instrument noise to the signals detected. Sample quantities and measurement times are several times lower for NMR spectrometers equipped with such probes than they are for spectrometers that operate with conventional room-temperature probes.

The molecular weight limit for NMR structure determination will probably increase twofold or more in the next few years. It has been discovered that both the sensitivity and the resolution of certain NMR measurements crucial for characterizing proteins, specifically those that involve protein backbone nitrogen sites and aromatic side-chain carbon sites, can be greatly enhanced by exploiting the cancellation between nuclear magnetic dipole-dipole interactions and anisotropic chemical shifts that occur at fields near 20 T. This cancellation greatly reduces certain nuclear spin relaxation rates (decoherence rates), which leads to sharper NMR lines and reduced signal losses in multidimensional NMR experiments. New radio-frequency pulse sequence techniques such as TROSY take advantage of this effect. When TROSY techniques are combined with other recent developments such as protein chemical shift databases, residual dipolar couplings in aligned systems, and cross-correlated relaxation effects, full-structure determinations should become possible for proteins having molecular weights around 80 kDa.

It should be emphasized that although the phenomenon that makes TROSY possible has been understood for years, its significance for biomolecular NMR was entirely unanticipated and was discovered only as a result of observations made once NMR spectrometers operating at fields near 20 T became available. The lesson of this experience is that the development of magnets that operate at higher fields can foster progress by leading unexpectedly to new ideas, new observations, and new approaches that promote scientific progress.

Solid-State NMR

The molecular NMR spectroscopy discussed above is done on substances dissolved in water or some other liquid. A different type of NMR, called solid-state NMR, is used to characterize solids of all kinds, including insoluble aggregates or assemblies of biopolymers—for example, the amyloid deposits associated with Alzheimer's disease and proteins embedded in biological membranes.

Solid-state NMR is a well-established discipline in chemistry. It has been particularly important for investigating the molecular structures of polycrystalline and noncrystalline inorganic materials, such as the zeolites used as industrial catalysts and the glasses used as stabilization media for nuclear waste. For example, the distributions of local structures and bonding geometries within glasses, determined from solid-state NMR measurements, provide critical tests of theories of glass structure. Solid-state NMR is much less widely used for biological research than solution NMR and has been slower to develop because the resolution and sensitivity of most existing solid-state NMR spectrometers are too low to permit routine determination of the structures of macromolecules in the solid state. One reason resolution is poor in solid samples is that the ^{1}H NMR signals on which structure determinations largely depend are broadened by strong nuclear magnetic dipole-dipole interactions that in solution are normally averaged to zero by molecular tumbling, as noted earlier. A solid-state technique called magic-angle spinning removes broadening due to dipole-dipole and other interactions to lowest order, but higher-order effects remain, which still broaden solid-state NMR resonances. Nevertheless, techniques are being developed for obtaining structural constraints for complex biochemical systems by solid-state methods.

Higher-field magnets would help alleviate the line broadening problem in solid-state NMR because the higher-order effects of dipole-dipole interactions, which are field-independent, become less important relative to the chemical shift differences as the magnetic field increases. It is also worth noting that the magnetic field homogeneity and stability requirements for solid-state measurements are not as stringent as for biological NMR measurements. A 30- to 50-T resistive or hybrid magnet, which was unusable for conventional multidimensional biological NMR, could be very valuable for solid-state NMR of biological or inorganic materials.

Improvements in field strength could lead improved performance in other ways. Many of the materials studied by solid-state NMR contain quadrupolar nuclei such as ^{27}Al, ^{17}O, ^{23}Na, and ^{67}Zn. NMR measurements on quadrupolar nuclei at low fields are notoriously difficult because, even with magic-angle spinning, the NMR line from a single site may be extremely broad (~10^4 Hz) as a result of orientation-dependent, second-order quadrupole shifts. In complex structures with many inequivalent sites, the NMR lines of individual sites are commonly unresolvable. However, the second-order quadrupole shifts are inversely proportional to the applied field, H_0, while the chemical shifts that distinguish one site from another are directly proportional to H_0. Thus, the resolution of such NMR spectrum should increase quadratically with H_0. The line-narrowing that accompanies the reduction in second-order quadrupole shifts should lead to an even stronger increase in sensitivity at higher fields. Solid-state NMR measurements on inorganic materials therefore stand to benefit enormously from the availability of higher field magnets.

NMR in Condensed-Matter Physics

Since the discovery of high-T_c superconductivity in cuprate materials in 1986, there has been a resurgence of interest in NMR in the condensed-matter and materials physics community. Measurements of the dependence of ^{63}Cu, ^{65}Cu, ^{17}O, and ^{89}Y NMR frequencies, spin-lattice relaxation rates, and electron-mediated couplings and the dependences of these quantities on temperature and doping have provided important information about the electronic structure of high-T_c ceramics, the nature of antiferromagnetic electron spin fluctuations, and the nature and symmetry of the Cooper pair states that give rise to superconductivity in these materials. NMR measurements provided the first experimental evidence for spin-singlet, *d*-wave pairing in the superconducting state of high-T_c ceramics.

NMR has also contributed to our understanding of fullerenes (molecular forms of carbon, such as the buckyball, C_{60}) and the superconducting alkali fullerides, which were discovered in 1990-1991. NMR measurements have provided important information about the molecular dynamics, phase diagrams, and electronic properties of these materials. ^{13}C NMR spectra and spin-lattice relaxation measurements were the first to reveal the rapid molecular rotations in solid C_{60} and permitted the determination of kinetic parameters that affect sound velocity and thermal conductivity in these materials. NMR measurements on superconducting alkali fullerides provided early estimates of the electronic density of states, superconducting state energy gap, and magnetic penetration length.

Beginning in 1994, optically pumped NMR techniques were used for the first time to study the properties of 2D electron systems in GaAs/AlGaAs quantum well structures. Optical pumping, in which optical excitation of electron-hole pairs in GaAs layers leads to enhanced nuclear spin polarizations and enhanced NMR signals, permits measurements on semiconductor thin films and quantum wells containing $\sim 10^{16}$ nuclei. The NMR data provided the first experimental evidence for skyrmion states in 2D electron systems in the extreme quantum limit of low temperatures and high fields—that is, electronic states with effective spins greater than 2. The optically pumped NMR data, which demonstrated the existence of skyrmions near a Landau level filling factor $\nu = 1$, stimulated many subsequent theoretical and experimental studies of skyrmions in 2D electron systems and sparked renewed interest in the electron spin properties of such systems in the FQHE regime.

These examples illustrate the potential of NMR to provide information about strongly interacting electron systems that is complementary to the information obtained from the measurements of transport properties, susceptibilities, and optical properties that are commonly performed in solid-state physics. These NMR measurements depend on the availability of stable and homogeneous high magnetic fields, but as in the case of the inorganic materials discussed above, the stability and homogeneity requirements are not as stringent as in most biological NMR applications.

The development of higher fields will generally increase the sensitivity of NMR measurements in condensed-matter and materials physics. It could, for example, make NMR studies of semiconductor heterostructures possible without optical pumping. In addition, the availability of higher fields with adequate homogeneity and stability will allow intrinsically high-field phenomena to be studied by NMR. NMR measurements on skyrmion states and fractional quantum Hall states in GaAs/AlGaAs quantum wells are one example. Another impressive example is provided by recent studies of field-induced magnetic ordering in quasi-2D transition metal oxides, carried out at the Grenoble High Magnetic Field Laboratory, in France, in which ^{63}Cu and ^{65}Cu NMR spectra were recorded as a function of field, up to 28 T and at 35 mK. These studies revealed drastic and unanticipated changes in the NMR spectra beginning at approximately 27 T, from which the field-induced magnetic superlattice structure could be inferred.

The new fields of spintronics and quantum computation are two additional areas in which NMR measurements could have an impact. Relatively simple solution NMR experiments, quite analogous to the pulse sequence techniques commonly employed in biological NMR, have already been carried out to demonstrate the basic principles of quantum computation. In these experiments, nuclear spins represent qubits, and radio-frequency pulse sequences are used to implement quantum logic gates. Future implementations of quantum computation are likely to be based on microfabricated semiconductor devices, in which the qubits may be quantum-confined electron spins, nuclear spins, optical transitions, or some combination of these. High magnetic fields and NMR-related techniques will play an important role because the ability to address individual qubits and minimize the decoherence of quantum information will increase with increasing field.

Magnetic Resonance Imaging

MRI is a noninvasive technique for determining the spatial distribution of nuclear spins in samples. In biological applications, the spin distribution studied is usually that of ^{1}H (^{129}Xe, ^{3}He, and ^{13}C are also used), and since water is by far the most abundant hydrogen-containing substance in tissue, most biological MRI images show how water is distributed in the sample, which may be a live human being. In MRI experiments, samples are placed in magnetic fields that vary in strength across the sample volume in a well-defined way. All of the nuclear spins in the sample are then perturbed using radio-frequency pulses, and the radiation emitted by the sample is detected. Provided the point-to-point variation in field strength in the sample is large compared with chemical shift differences, differences in the resonant frequencies of spins will encode differences in location, not differences in chemical state.

MRI is now routinely used by clinical radiologists instead of x rays to obtain images of soft tissue in the human body and is the preferred method for diagnosing brain tumors, multiple sclerosis, and other neurological and spinal conditions and some forms of cancer. Recent work suggests that it may be possible to determine the metastatic potential of a tumor accurately from a spectroscopic image in which the chemical composition of a tumor and its location and size are determined by MRI. MRI is also being developed as a method for determining in advance the precise location and depth of the incisions that should be made during surgery to minimize tissue damage and patient risks.

An extremely important and entirely unanticipated advance has been the development in the past 15 years of MRI methods for imaging brain activity. This technique, called fMRI, is revolutionizing the neurosciences. Psychologists use fMRI to determine the regions of the brain that participate in specific thought processes and emotional responses and to develop new classifications for cognitive tasks based on the regions of the brain they activate. fMRI may become a diagnostic tool for psychiatry.

The sensitivity of MRI images to brain activity is believed to derive from the changes in local blood flow and oxygen consumption when a specific area of the brain increases (or decreases) its level of function. These physiological changes alter the local concentration of deoxyhemoglobin, a paramagnetic molecule. The relaxation rates of the nuclear spins change in response because changes in deoxyhemoglobin concentration affect the local magnetic field homogeneity. It is easy to operate MRI instruments so that fast relaxing spins are distinguished from slow relaxing spins.

MRI instruments currently account for the vast majority of superconducting magnets sold commercially. The field strengths of the magnets used in the instruments employed to image the human body are lower than those for ordinary NMR spectroscopy, but the sample volumes they accommodate are much larger. Field strengths are up to 9.4 T (in a small number of research groups), and bore diameters of 90 cm are typical.

Higher fields improve the sensitivity of MRI instruments just as they improve the sensitivity of NMR spectrometers, and in this case improved sensitivity translates into improved spatial resolution. Furthermore, in fMRI experiments, small changes in signal intensities must be detected following specific stimuli. The information content of fMRI data is thus strongly limited by the signal-to-noise ratio, so higher fields are beneficial here, too. However, issues related to patient safety may limit what can be done in this area as much as issues related to the construction of MRI instruments operating at higher fields.

Magnetic resonance "microimaging" experiments are carried out on small laboratory animals or cell cultures using instruments equipped with magnets having much smaller bores than those used in the clinic. To date, microimaging experiments have been done using magnetic fields of up to 17.6 T. The spatial resolution achieved is on the order of 10 microns, and is again limited by the signal-to-noise ratio. The imaging of amyloid plaques in transgenic mouse models of Alzheimer's disease has been reported. Techniques are being developed to track movements of individual cells that are appropriately labeled with paramagnetic contrast agents. Development of compounds that can be used as cell-type-specific and protein-specific contrast agents for MRI and microimaging is an active area of chemical research. Microimaging at even higher fields and with new contrast agents may make it possible to follow cell movements in a developing tissue, changes in gene expression, and movements of subcellular structures noninvasively and in real time in live laboratory animals. These innovations could make important new classes of experiments possible for developmental biologists and endocrinologists, among others.

Prospects for Improvements with Still Higher Fields

Several ways in which increases in magnetic fields can improve NMR performance have already been mentioned. Over the past 20 years, the development of higher field magnets for solution NMR has been an incremental process. NMR spectrometers operating at ^{1}H NMR frequencies of 270 MHz, 360 MHz, 400 MHz, 500 MHz, 600 MHz, 750 MHz, 800 MHz, and 900 MHz have been developed in sequence, with each step requiring 2-4 years to accomplish (see Figures 2.6 and 2.7). Most of these steps took place in industry, but with an increasing degree of academic research collaboration. For example, one of the most recent 900-MHz spectrometers to come on line was developed almost entirely at NHMFL. It is likely that instruments operating at still higher fields will one day be developed in a similarly incremental way and that this will gradually increase the complexity of the systems that can be characterized successfully by NMR. It is important to realize, however, that although each step in the process has been small, the cumulative effect has been large. The 800- and 900-MHz NMR spectrometers available today are vastly superior to the 360-MHz spectrometers that came on the market 20 years ago in almost every respect.

If NMR spectrometers with stable, homogeneous fields of 30 T or higher (1.3 GHz or higher) were to become available in the future, the impact on both biological and nonbiological users would be considerable. The increased sensitivity and resolution of such an instrument would probably allow using solution NMR to determine the structure of proteins two to five times larger than the proteins whose structures can be determined today. The tendency of macromolecules to align in magnetic fields, which is detectable but quite small in today's NMR spectrometers, would become more significant (scaling as the square of the field strength), enabling new approaches to structure determination through anisotropic nuclear spin interactions. Biological solid-state NMR measurements would be qualitatively transformed, principally because ^{1}H NMR signals would become well enough resolved to begin to be as useful and informative as they already are in solution NMR. Progress would be made in solving the general problem of membrane protein structures. Recent studies of relatively small model proteins in microcrystalline form have demonstrated the feasibility of full structure determination by solid-state NMR. The improved spectral resolution at higher fields will facilitate the extension of these results to larger membrane proteins. Solid-state NMR spectroscopy of inorganic materials would improve dramatically as second-order quadrupole effects become small spectral perturbations rather than the dominant feature in NMR spectra of quadrupolar nuclei.

As mentioned earlier, certain types of NMR have less stringent requirements for magnet homogeneity and stability than solution NMR. In particular, many biological and nonbiological solid-state NMR measurements can be performed in fields with about 1 ppm homogeneity over 0.1 cm^3 (rather than 1 ppb over 1 cm^3). Furthermore, magnet instabilities are manageable if the field drift can be calibrated or monitored during experiments. Provided drift rates are not too great, it

FIGURE 2.6 The magnet of the 900-MHz NMR spectrometer in Pacific Northwest National Laboratory's Environmental Molecular Sciences Laboratory. This picture illustrates some of the issues associated with high-field magnets. This 21-T magnet was manufactured by Oxford and has a 63-mm room-temperature bore and a stored energy of 27 MJ; it is sited inside a cylindrically shielded enclosure 24 ft in diameter and extends 15 ft above and below the main laboratory floor (the magnet itself is 8 ft in diameter and 21 ft tall). The center of the magnetic field is located at exactly floor level to preserve the desired symmetry of the magnet and shield. The person standing next to the magnet is just over 6 ft tall. Photo courtesy of William R. Wiley Environmental Molecular Sciences Laboratory, Pacific Northwest National Laboratory.

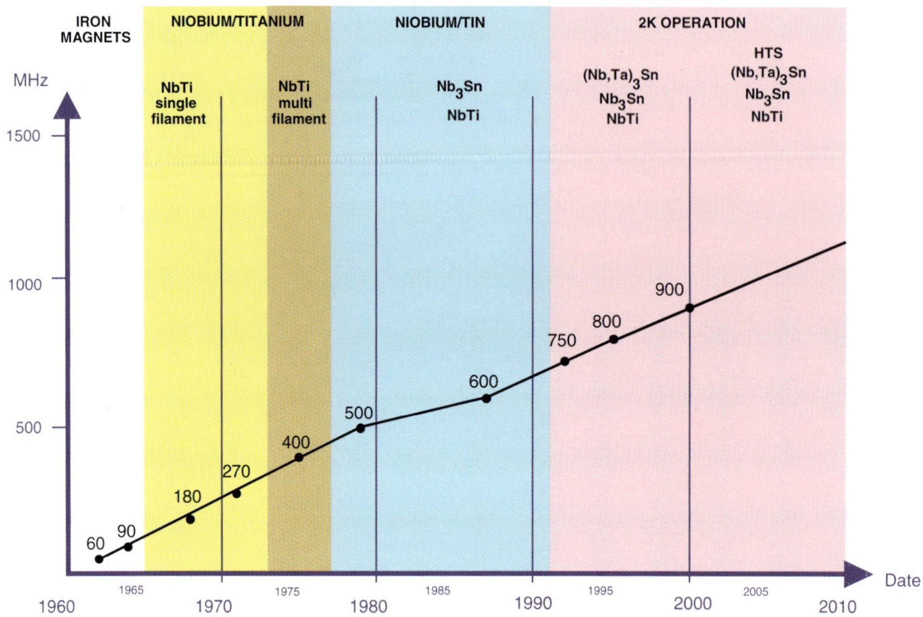

FIGURE 2.7 The growth of the field strength of NMR spectrometers in common use. The types of superconducting wire used for the highest-field NMR magnets in each era are indicated at the top of the figure. Before 1992, magnets operated at the temperature of liquid helium (about 4 K). Since then, the highest-field magnets have required pumped-helium cooling systems that provide even lower temperatures (about 2 K). Current technology (Nb-Ti, Nb_3Sn, and $(Nb,Ta)_3Sn$) may enable 1-GHz operation, but new high-field conductors based on MgB_2 or HTS materials will have to be developed for higher frequencies. (The committee notes that a 600-MHz NMR magnet was introduced in 1978 at Carnegie Mellon University; the magnet was constructed of Nb_3Sn tape and was nonpersistent, setting it apart from the more heavily commercialized NMR magnets. As a proof-of-principle, this design may have important value for the future.) Figure courtesy of Bruker Biospin, Inc.

is possible to correct data for magnet drift either in real time or after the fact. Thus, a 30- to 50-T magnet for NMR, with 1 ppm homogeneity over 0.1 cm^3 and 1 ppm/min drift specifications, would be valuable even if it could not be used for high-resolution NMR of proteins in solution.

For any such a magnet to achieve its full potential, however, ancillary equipment would have to be developed. In particular, NMR probe and cold-probe circuitry that permits double- and triple-resonance experiments at ^1H NMR frequencies of 1.3 GHz and above will have to be developed and incorporated into the probes, including those with high-speed, magic-angle spinning capabilities. In addition, capabilities for both low-temperature and high-temperature measurements must be part of the package as well as radio-frequency power amplifiers that produce ~500-W pulses at the NMR frequencies of hydrogen and other relevant nuclei. Finally, the spectrometer console, which controls the execution of pulse sequences and the acquisition of signals, must have the full capabilities of a modern NMR instrument.

Strategic Considerations for Higher Field NMR

The committee believes that the construction of 1.3-GHz NMR spectrometers is a scientifically justifiable objective. Its endorsement is strongly influenced by the fact that in the past, NMR spectroscopy has shown an astonishing capacity to reinvent itself in productive but unanticipated ways in response to improvements in instrumentation. However, the committee realizes that this enterprise will present major challenges for the agencies that support research dependent on high-field NMR spectroscopy, the people who manufacture these instruments, and the user community. The highest-field NMR spectrometers available today, which operate at 900 MHz, cost approximately $5 million each, and the cost of a 1.3-GHz spectrometer, assuming that it can be built, will certainly be much higher, perhaps as much as $20 million. Thus, unlike the number of 600-MHz NMR spectrometers available in the United States today, for example, the number of 1.3-GHz NMR spectrometers will never be large, and demand for access to them is certain to exceed supply. Two conclusions follow:

- The technical challenges that surround the construction of the magnets required for these spectrometers are so great and the potential market for them is so small that manufacturers are unlikely to undertake their construction without at least some direct public support. Thus, in addition to playing their traditional role as the main purchaser of the product, federal agencies interested in high-field NMR research will have to shoulder some of the risk up front or these instruments will not be built.
- Although the NMR data obtained with a 1.3-GHz spectrometer is expected to be of substantially better quality than that obtained with current spectrometers, which should allow new problems to be addressed, the throughput of these instruments might only be twice that of current 900-MHz instruments. Thus the NMR community will need to address the challenge of providing fair access to these spectrometers for all interested parties while at the same time ensuring that only experiments that absolutely require the highest available fields are performed on them. Given that the NMR community is used to running experiments on locally controlled instruments rather than sharing centralized facilities, some social engineering may be required to ensure a sensible outcome.

Ion Cyclotron Resonance

The phenomenon of ICR can be used to determine atomic and molecular masses. The technique is called ion cyclotron resonance mass spectroscopy (ICRMS). In an ICRMS instrument, modest numbers of ionized molecules are introduced into an ion trap in the middle of a homogeneous magnetic field. The trajectories of such ions are circular, viewed along the direction of the field, and the frequencies at which they orbit are determined by their mass-to-charge (m/z) ratios, and are proportional to magnetic field strength. The radii of ion orbits at the time of injection are small, and since the ionized molecules in the injected sample are randomly distributed around those orbits, viewed from the outside, the circulation of these ions has no net electrical effect. Ions can be forced to enter orbits of higher radius by the application of radio frequency pulses of the appropriate frequency, duration, and polarity, and the coherent orbital motions of the ion clouds that result induce oscillating voltages in detection circuitry that surrounds the ion traps, from which m/z values can be determined. (The similarities between ICRMS and NMR experiments are many and obvious.)

Because the lifetimes of the coherent cyclotron motion of such clouds, which are limited by collisions and by magnetic field inhomogeneities, can be quite long, m/z values can be determined with remarkable precision (~1 ppm) by ICRMS. The molecular weights of the components of mixtures containing thousands of chemically distinct ions can be determined simultaneously by ICRMS. The mass accuracy can be less than that of a single electron, so that chemical compounds with the same nominal molecular weight but different elemental compositions can be distinguished by ICRMS.

ICRMS is rapidly growing in importance. One application is in the field of proteomics, where the goal is to determine the full array of proteins expressed within a given cell type or tissue (and the posttranslational modifications of these proteins)—for example, as a function of exposure to hormones, drugs, or other bioactive compounds; as a function of developmental stage; or as a function of disease state. These measurements can be done with ICRMS by extracting proteins from cells or tissue, fragmenting them into shorter peptide segments, and then determining the masses of all fragments. The sensitivity of ICRMS is remarkably high: About 100 ions of the same species can generate a detectable signal. Thus, it is entirely compatible with the usual scales of molecular biology experiments. Because of the sensitivity and resolution of ICRMS spectra, ICRMS methods may be employed in the future to establish biomarkers for disease states in humans. ICRMS is also a tool of unparalleled power for determining the composition of complex mixtures such as crude oil.

ICRMS benefits from high magnetic field. Commercial ICRMS spectrometers currently operate at fields up to 12 T, and instruments up to 14.5 T are under construction. Provided that field homogeneity is not limiting, resolving power increases linearly with field, and upper mass limit and dynamic range improve quadratically. Homogeneity requirements are not as stringent as in high-resolution NMR because the orbital motions of ions tend to average out field inhomogeneities. Thus, magnets with about 10 ppm homogeneity and about 10 ppm short-term stability are suitable for ICRMS. High-field resistive magnets could conceivably be used.

ICRMS is still in its infancy, and it has benefited mightily from the many techniques that have been developed in recent years for preparing molecular ions for mass spectrometric analysis. New applications are being discovered every day, and it is likely that before long, every research university will feel it must have ICRMS spectrometers on its campus. As it happens, ICRMS is one of the big success stories at NHMFL, whose Center for Interdisciplinary Magnetic Resonance in Tallahassee, Florida, has been an important locus for the development of ICRMS ever since it was founded.

Electron Paramagnetic Resonance

The phenomenon of EPR was discovered in bulk matter by E.K. Zavoisky in 1944, before the initial observations of NMR. EPR and NMR spectroscopy are similar in many respects; both examine the response of a sample exposed to radiofrequency radiation in a magnetic field. In EPR, the species detected are unpaired electrons. In NMR, it is nuclei that have nonzero spins. The technical differences between the two kinds of spectroscopy derive mainly from the difference in gyromagnetic ratio between electrons and nuclei. In a 1-T field, for example, the resonant frequency of a simple organic free radical is about 28 GHz, but the resonant frequency for protons is only 43 MHz. The wavelength of 28 GHz radiation is around 1 cm, which implies that a 1-T EPR spectrometer must be equipped with microwave electronics, a constraint that limits what can be done. Microwave devices tend to be relatively narrow-band, so EPR spectra are usually collected by setting the electronics of the spectrometer to a single frequency and then sweeping the field strength of its magnet over a wide range so that the different paramagnetic species in the sample can be brought into resonance one at a time. EPR spectrometers available commercially today operate at 95 or 130 GHz and have 3.5- to 5-T magnets.

As is the case for NMR and ICRMS, in principle, the performance of EPR spectrometers improves with increasing magnetic field, but there are limitations. If an EPR spectrometer were built around a 21-T magnet, i.e., a state-of-the-art NMR magnet, the resonant frequencies for simple organic free radicals would rise to about 588 GHz and wavelengths fall to about 500 μm. For some of the metal centers (in metalloproteins and surface-catalytic metals) interesting to chemists and materials scientists, the frequencies might be even higher, about 6 THz, and the wavelengths even shorter, about 50 μm. Radiation having these characteristics lies in an awkward part of the electromagnetic spectrum, between microwaves and the far infrared; it is hard to work with.

The development of an integrated spectrometer operating at 21 T would present major challenges, but the rewards could be commensurate. New radiation sources, phase shifters, resonators, digital switches for pulse generation, power amplifiers, and preamplifiers would all have to be devised. However, if they were, it would become possible to do EPR experiments equivalent to the multidimensional, double-resonance experiments now routine in NMR at all field strengths and in EPR only at low field (9 GHz). The impact would be transformative. The spin Hamiltonian explored by NMR spectroscopy is simple to deal with theoretically because each of its terms is three to six orders of magnitude smaller than the dominating Zeeman interaction. It is different for EPR spectroscopy, because many of the internal interactions that must be taken into account are comparable to, if not larger than, the electron Zeeman interaction. Thus, electron spin Hamiltonians often depend on a dozen or more parameters that cannot easily be determined independently. At high field, EPR spectroscopy will approach the limit where these parameters can be measured accurately from first-order spectra. In addition, many interesting metal-centered species important in catalysis and materials have EPR transitions that can be measured only at high fields because of the large internal fields inherent in the metal ion or the material containing them.

Specialized, one-of-a-kind continuous-wave EPR spectrometers have been used in Europe, Israel, and NHMFL to demonstrate that metal centers of the kind just described can indeed be observed at high field in catalysts and optoelectronic and magnetic materials, demonstrating the potential of high-field EPR. It is time to develop the full range of pulsed EPR capabilities for high field so that the measurements now made routinely on simple organic free radicals and some metal centers at conventional EPR fields can be done on systems that require high field. Development of pulsed EPR spectrometers that operate at 15 T and beyond is already under way in the university and government labs of Germany, the Netherlands, and Italy. The United States should not allow itself to fall behind in this area.

Importance of Ancillary Technological Development

As has been pointed out repeatedly above, technological and methodological developments have led to constant advancement of the capability of magnet-based measurement devices and a concomitant increase in their impact on many areas of science. The development of higher field magnets has played a central role in these advances to be sure, but the importance of developments in other areas of technology should not be forgotten. In fact, without these advances in ancillary technologies, the full potential of higher-field magnets would not have been realized, and, given the expense of high-field magnets, investments in technologies needed to optimize their utility makes good economic—as well as scientific—sense. While federal funding for the application of existing technology and methods to specific scientific problems has generally been good, federal funding for the development of novel technology and methodology has been inadequate.

NMR and MRI instrument manufacturers have done a good job of advancing the ancillary technology relevant to these techniques when relatively large commercial markets for their products justified their doing so. For example, the recent development of cryogenically cooled probes for high-field solution NMR, which enhance sensitivity severalfold, was carried out entirely by instrument manufacturers, based on their expectation about the market for them. However, there are many other areas where technological advances are sorely needed but where the commercial market is not large enough to attract the attention of instrument manufacturers. Several examples follow.

- Biological solid-state NMR measurements on integral membrane proteins, which, in principle, could be used to determine the structures of receptor-bound hormones, neurotransmitters, and other biological signaling molecules, are severely limited by the availability of appropriate samples. It is often very difficult to make such materials in quantities sufficient for the instrumentation now available. This problem would be greatly alleviated by the development of magic-angle spinning probes for solid-state NMR that operate at 10-30 K. Further sensitivity enhancements might be obtained by techniques such as dynamic nuclear polarization (DNP) and optical pumping. Recent DNP experiments, in which paramagnetic compounds are added to frozen solutions to permit cross-relaxation between electron spins and nuclear spins during microwave irradiation of the electron spins, suggest that signal enhancements of two orders of magnitude may be achievable quite generally with further developments in microwave technology, solid-state NMR probe design, and paramagnetic reagents.
- In the area of MRI, the advent of whole-body imaging systems operating at 8 T and above creates new problems in detector coil design, largely because the wavelengths of the radio-frequency radiation detected are comparable to anatomical dimensions, making the amplitudes of the radio-frequency fields within the body highly nonuniform. Optimal coils for high-field MRI will probably not be developed unless a significant research program is undertaken by groups outside the commercial sector, and this is unlikely to

happen without federal support. Similarly, MRI microscopy at the highest available fields, which could be used for noninvasive imaging at the cell or organelle level within model organisms, would benefit significantly from the development of improved microcoils and field gradient coils.

- Hybrid magnets with fields greater than 40 T may not have the homogeneity and stability required for biological NMR, but they do have considerable potential for NMR studies of inorganic materials and phenomena in condensed-matter physics. Realization of this potential will require the design and construction of appropriate NMR probes with high- and low-temperature capabilities and the development of field stabilization methods to extract the maximum possible spectral resolution and sensitivity.
- Finally, perhaps more than any other class of experimental techniques now in common use, NMR and MRI have both benefited from methodological advances that are purely concept-based, as opposed to equipment-based. Physical, chemical, and biological scientists have benefited enormously from the design of radio-frequency pulse sequences that excite and manipulate nuclear magnetic moments in new ways. Developments in pulse sequence methodology have usually resulted from new insights into the way nuclear magnetic moments evolve in external and internal magnetic fields and from new ways to describe these evolutions mathematically. Because higher fields significantly change the relative strengths of the interactions that determine how nuclear magnetic moments evolve in response to radio-frequency pulse sequences, improvements in pulse sequences will be required if the field is to take full advantage of high-field magnet development.

Projects aimed at improving NMR and MRI capabilities, from which large numbers of investigators are certain to benefit, are surely as worthy of support as the specific projects based on current NMR and MRI techniques. A strong program requires a healthy balance between research aimed at improving current capabilities and research that exploits current capabilities.

OTHER SCIENTIFIC USES OF HIGH-FIELD MAGNETS

High-field magnets are critical components of instruments used in several research areas that so far have been mentioned only in passing, most notably high-energy and nuclear physics, plasma science, and fusion energy research. Advances in high-energy physics have long been strongly coupled to developments in magnet technology, because increases in magnet performance have been required for the construction of ever more power particle accelerators and particle detectors.

The U.S. high-energy physics community established an early dominance in superconducting accelerator technology by constructing the Tevatron at Fermi National Accelerator Laboratory (FNAL), which is still the world's highest energy accelerator. Its 4-mile circumference ring includes 1,000 Nb-Ti superconducting magnets supplying a field of 4.4 T. It began operation in 1983 and has been upgraded over time to reach an energy of 1 TeV. An even larger accelerator is the Large Hadron Collider (LHC) at CERN, designed to collide proton beams at 14 TeV. Still under construction, it will use a dual-aperture beamline 27 km in circumference, which includes 5,000 Nb-Ti superconducting magnets. These magnets will operate at 1.9 K with a peak field of 8-9 T. The expected start-up date is 2007. Because Nb_3Sn is brittle, its use requires special magnet construction techniques that are more costly than those for the tough and ductile Nb-Ti. Thus, unless the higher-field performance is required, Nb-Ti is generally employed, as it was for the LHC, where it was found easier to design with Nb-Ti than Nb_3Sn. A prototype 16-T dipole magnet for the next generation of accelerator dipole magnets (for proton machines) was demonstrated in 2003 at LBNL; these magnets will require Nb_3Sn, MgB_2, or an HTS conductor.

High-energy physics magnets are predominately dipoles, quadrupoles, and higher order configurations, but although their field and force distributions are very different from the solenoids used for most other purposes, the stresses and other engineering problems confronted in their construction are similar. The Tevatron and the LHC achieve their goals using huge quantities of Nb-Ti magnets, but it is unlikely that Nb-Ti magnets will suffice for the large circular accelerators that might succeed the LHC. The field strengths attainable by Nb-Ti magnets will simply not be enough. For this reason, the U.S. high-energy physics community has established a research program to develop Nb_3Sn-based dipoles and quadrupoles. LBNL has achieved a 16-T peak field in a small dipole prototype coil using an advanced, very high critical current density Nb_3Sn wire. Other laboratory partners in the DOE high-energy magnet technology effort include Fermilab and Brookhaven National Laboratory. Fields up to 20 T may be achievable with further development, but this will require a focused program devoted to the commercial production of large quantities of very high current density Nb_3Sn wire. Because the different magnet user communities have been isolated from one another, developments in accelerator magnet technology have not had a broad impact outside the field. Recently, however, researchers at FNAL and LBNL have begun to form partnerships with fusion science magnet designers and superconducting materials experts outside the traditional laboratory centers.

Research into the development of fusion as a future energy source has been going on around the world for years. Fusion devices are operating in Europe,

Russia, and Japan, and new ones are being constructed in South Korea, China, and India. All the fusion devices being used or being considered require very large superconducting magnets. The United States is now operating only one superconducting fusion device, known as the Levitated Dipole Experiment. However, it has had an extensive superconducting magnet development program under way since the 1970s. In the 1980s the extremely large superconducting mirror machine MFTF-B was started up at Lawrence Livermore National Laboratory, but it never went into full operation for reasons unrelated to magnet technology.

Fusion magnets come in many shapes and sizes, including solenoids, toroids, and helical coils. The devices presently in operation use Nb-Ti magnets, but newer machines will use Nb_3Sn magnets. The largest project now being planned is the International Thermonuclear Experimental Reactor (ITER), which will be a collaboration between the United States, Europe, Russia, Japan, China, South Korea, and India. The device will cost more than $5 billion and is scheduled to be constructed over 8 years beginning in 2006. This machine will require the commercial production of about 500 tons of high-quality Nb_3Sn superconductor over a several-year period, a more than 10-fold increase in world production of Nb_3Sn.

During the 1990s the parties involved in ITER made several large-scale prototype superconducting magnets. The largest of these was the Central Solenoid Model Coil, built jointly by the United States and Japan. Its coil has an inner diameter of 1.6 m, an operating current of 46,000 A, a peak field of 13 T, and a stored energy of 640 MJ. It can be operated as a DC magnet or ramped from zero field to 13 T in 8 s without quenching.

These examples suggest the broad utility and critical importance of high-field magnet science and technology in fields beyond condensed-matter physics, materials science, and magnetic resonance. Although the scale of application for high-field magnets in a high-energy particle accelerator is vastly different than that associated with the study of correlated-electron systems, both communities drive—and benefit from—general advances in high-field magnet technology. It is important to note that these disparate communities of users have not traditionally collaborated on magnet technology. Indeed, the recent convergence of the particle physics and fusion science magnet efforts was precipitated more by their shared source of funding (DOE's Office of Science) than by any overlap of ongoing research efforts. It is nevertheless the case that advances in magnet design, construction, and performance made by one community can significantly benefit other communities.

3

Technological Challenges and Opportunities for Developing Higher Fields

The design and construction of magnets that operate at high field is an art form that requires the balancing of many conflicting requirements. This chapter starts with a discussion of the nature and magnitude of the challenges that must be met and concludes with a description of the opportunities that now exist for improving magnet performance. A common thread in this discussion is the distinction between structural problems that limit what can be done and intrinsic material properties that often have not yet been exceeded, suggesting that it should be possible to build magnets that operate at fields significantly higher than any available today.

WHAT IS THE CHALLENGE?

Four different types of electromagnets are used to generate high fields: (1) resistive DC magnets, (2) (resistive) pulsed-field magnets, (3) superconducting DC magnets, and (4) hybrid magnets, combining both resistive and superconducting elements. The maximum fields obtainable from magnets of all four types have increased significantly over the past several decades, but the increases have come in increments of 10-20 percent rather than in large jumps (such as by factors of 10 or more), and it is important to understand why this is likely to remain the case (Figure 3.1 shows how record field strengths have increased with time).

The reason advances in maximum field strength are hard to obtain is that, everything else being equal, both the energy stored in the field of a magnet and the

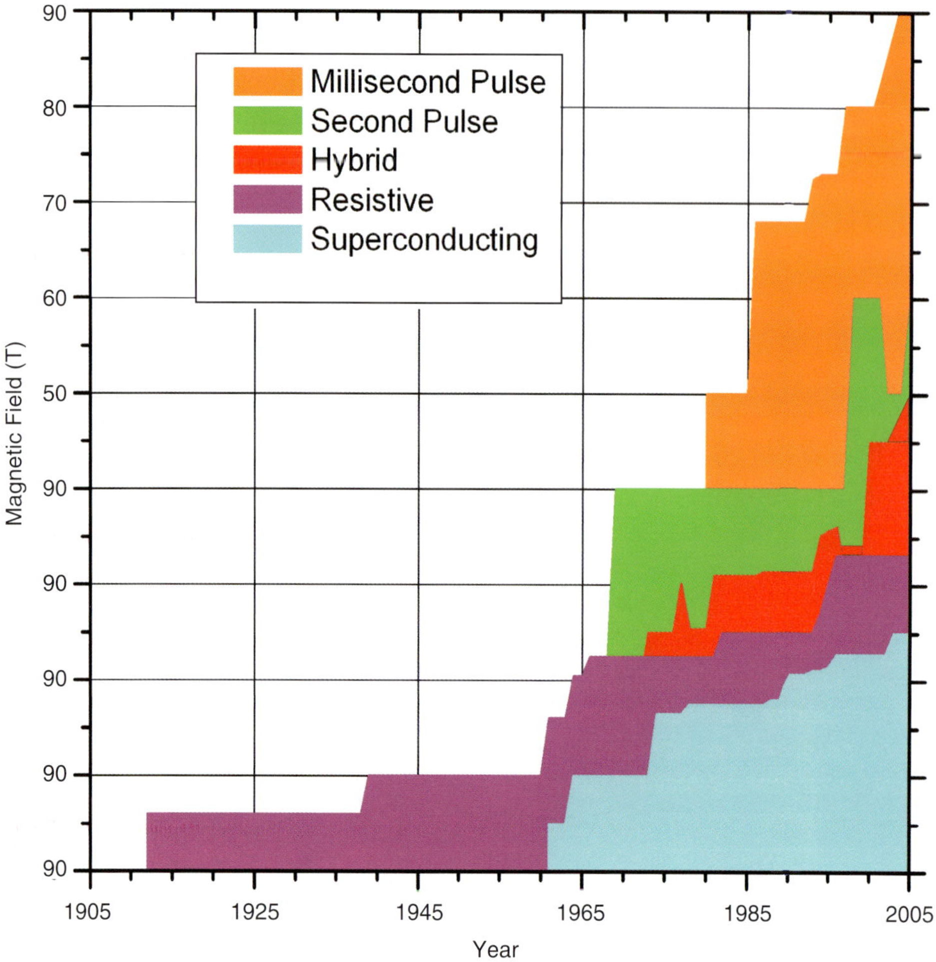

FIGURE 3.1 Available maximum field strength for nondestructive magnets at facilities around the world as a function of year. The maximum fields delivered by different types of magnets are indicated separately. Figure courtesy of G. Boebinger, National High Magnetic Field Laboratory.

stresses it experiences increase as the square of the field strength. Thus, the higher the field at which a magnet operates, the more complicated the stress and energy management techniques that must be used to ensure its physical integrity and the safety of those who work with it. All aspects of magnet design are affected, including the choice of materials for electrical conductors and structural components, the electrical insulation systems employed, and so on. In the case of superconducting magnets, field stability and quench protection pose additional issues.

The magnitudes of the energies and stresses that must be confronted are best illustrated with some numerical examples. The energy stored in the field of the highest-field magnet for high-resolution solution NMR now operating, which is the 21.6-T superconducting NMR magnet at the National Institute for Materials Science (Japan), is 33 MJ. This is roughly the kinetic energy of a 40-metric-ton railroad locomotive moving at 150 kilometers per hour, and the magnet is designed so that it will dissipate all of that energy harmlessly in the event the windings of its magnet become nonsuperconducting for any reason. Magnetic stresses are easiest to work out for long solenoidal magnets, because the mechanical forces generated by the interaction of the current in the windings of a solenoid with its own magnetic field are similar to an internal hydrostatic pressure. If the central magnetic field is 6 T, the equivalent pressure is 140 atm, which is about the working pressure of the typical gas cylinder. If the central field is 10 T, the internal pressure is about 400 atm, which is approximately the yield strength of annealed copper at room temperature. A 1-GHz NMR magnet (23.5 T) must withstand the equivalent of an internal pressure of over 2,000 atm and thus must be built of materials having yield strengths significantly greater than that of copper. Similarly large stresses must be dealt with in nonsolenoidal high-field magnets.

The magnetic field of a high-field magnet is generated only by the current-carrying conductor in its coil, but the coil necessarily includes other components. For example, none of the materials used as conductors in high-field magnets today is strong enough to withstand the stresses generated, so all magnets include mechanical support systems, which contribute nothing to the field. Electrical insulation and a thermal management system are also required. The higher the field to be generated, the more non-current-carrying material a coil must include and the lower its bulk-average current density. However, the lower the bulk-average current density, the bigger (and more expensive) the magnet must be to deliver a specified field. Given these facts, it should not be surprising that the inner bores of magnets tend to fall as the maximum field at the windings increases (see Figure 3.2), which reduces the amount of field energy that must be managed. Further, as just explained, the overall winding pack current density also tends to fall as stored energy rises (see Figure 3.3). (Table 3.1 describes the magnets shown in Figures 3.2 and 3.3.) Other issues that designers must consider include joints for both resistive and superconducting magnets and AC loss, stability, and quench problems for superconducting magnets. Finally, it is generally the case that no single aspect of a high-field magnet design can be altered independent of the others, so even incremental improvements in magnet performance must be hard fought for.

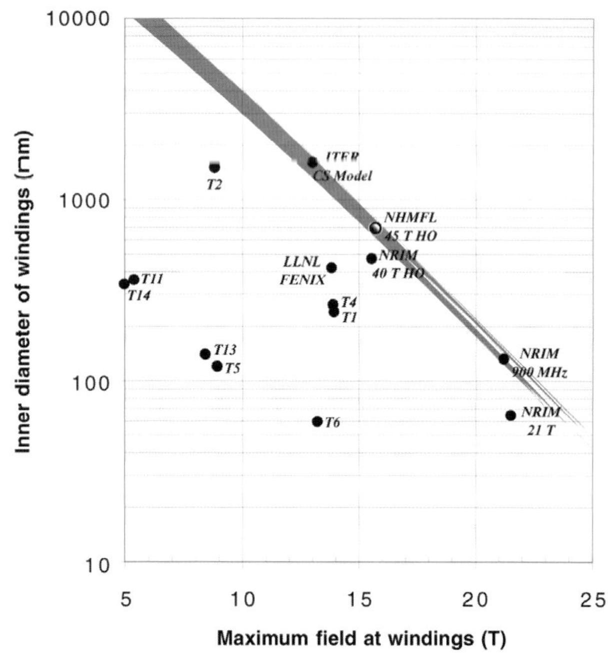

FIGURE 3.2 The stresses in solenoid scale with the square of the magnetic field, limiting the highest field magnets to small bore sizes and sample volumes. See Table 3.1 for legend. Figure courtesy of J.R. Miller, National High Magnetic Field Laboratory.

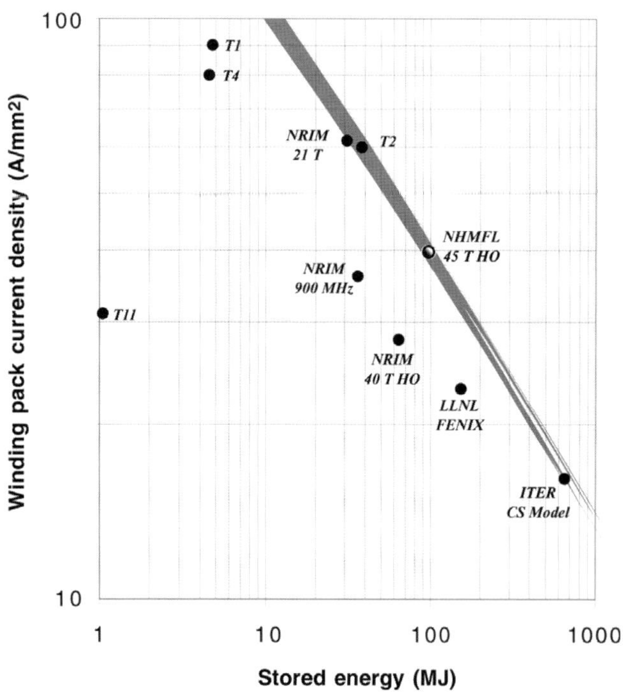

FIGURE 3.3 Overall winding pack current density decreases as the stored energy increases because increased amounts of structure, insulation, and thermal management are required to limit stresses and protect the magnet. See Table 3.1 for legend. Figure courtesy of J.R. Miller, National High Magnetic Field Laboratory.

TABLE 3.1 Description and Location of Magnets in Figures 3.2 and 3.3.

Item	Location	Description/Intended Use
T1[a]	Japan	DC solenoid/conductor testing
T2[a]	Japan	DC solenoid/conductor testing
T4[a]	Japan	DC solenoid/conductor testing
T5[a]	Japan	DC split solenoid/conductor testing
T6[a]	Japan	DC solenoid/conductor testing
T11[a]	Japan	DC pancake module/superconducting motors and energy storage (SMES) development
T13[a]	Japan	DC solenoid/SMES development
T14[a]	Japan	Pulsed pancake module/SMES development
NRIM 40 T HO[b]	Tsukuba, Japan	DC solenoid/research facility
NHMFL 45 T HO	Tallahassee, Florida	DC solenoid/research facility
FENIX[c]	Livermore, California	DC split solenoid/conductor testing
ITER CS model coil[d]	Japan Atomic Energy Research Institute	Pulsed solenoid/technology demonstration
NRIM 21 T[e]	Tsukuba, Japan	DC solenoid/research facility
NRIM 900 MHz[f]	Tsukuba, Japan	DC solenoid/research facility

NOTE: Includes a historical sample of foreign and domestic magnets. NRIM, National Research Institute for Metals; NHMFL, National High Magnetic Field Laboratory.

[a]O. Tsukamoto, S. Torii, T. Takao, N. Amemiya, S. Fukui, T. Hoshino, A. Ishiyama, A.Ninomiya, H. Yamaguchi, and T. Satow, Recent technical trends of superconducting magnets in Japan., *IEEE Transactions on Applied Superconductivity* 9(2), 547 (1999).
[b]H. Morita and S. Ito, Development of a 40 T Class Hybrid Magnet, *High Magnetic Fields: Applications, Generation, and Materials*, H. Schneider-Muntau, ed., Singapore, World Scientific, Singapore, 1997.
[c]D.S. Slack, R.E. Patrick, and J.R. Miller, FENIX: A test facility for ITER and other new superconducting magnets, *IEEE Transactions on Magnetics* 27(2), 1835 (1991).
[d]H. Tsuji, JAERI, Private communication, April 19, 2000.
[e]R. Hirose, T. Kamikado, O. Ozaki, M. Yoshikawa, T. Hase, M. Shimida, and Y. Kawate, 21.7 T superconducting magnet using $(Nb,Ti)_3Sn$ conductor with 14%-Sin Bronze, *Proceedings of the 15th International Conference on Magnet Technology*, Beijing, China, October 20-24, 1997.
[f]T. Kiyoshi, A. Sato, H. Wada, S. Hayashi, M. Shimada, and Y. Kawate, Development of 1 GHz superconducting NMR magnet at TML/NRIM, *IEEE Transactions Applied Superconductivity* 9(2), 559 (1999) and T. Kiyoshi, NRIM, private communication, March 6, 2000.

SOURCE: J.R. Miller, NHMFL.

RESISTIVE DC MAGNETS

Resistive DC electromagnets have been in use since the first half of the 19th century, and they can be as simple as a solenoid made of insulated copper wire. In addition to the force and energy problems described above, high-field resistive magnets present two unique challenges. First, they consume electric power in large quantities and hence need large DC power supplies. Second, most of the power they consume is converted into heat; thermal destruction is a serious problem.

Present Status

The most powerful resistive DC magnets operating today are magnets of a modified Bitter design that generate fields up to 33 T (see Figure 3.4). The field strengths of these magnets are limited primarily by power availability and by cooling issues, but also to some degree by the mechanical strength of materials in the assembly. For economic reasons, high-field DC magnets are found only at large

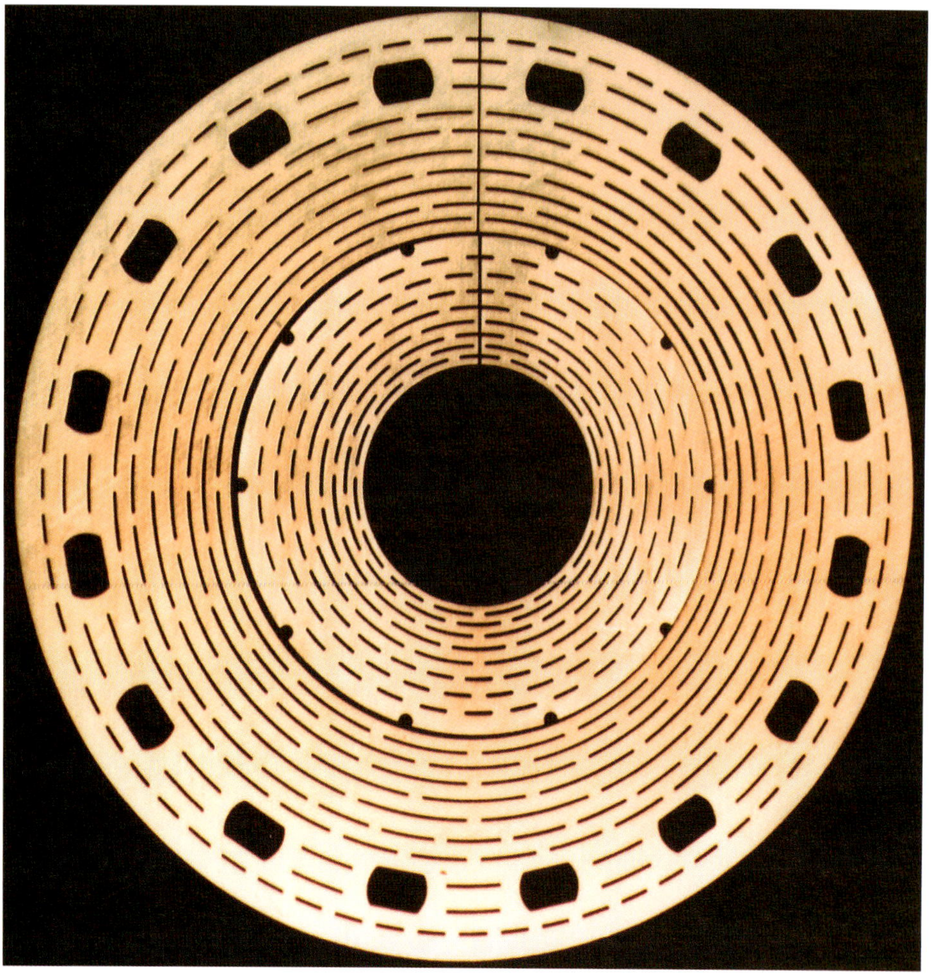

FIGURE 3.4 Typical plate from a (resistive) Bitter magnet. The small slots are for the circulation of cooling water. The ring of large holes is for axial prestress and accommodates structural tie rods. Image courtesy of National High Magnetic Field Laboratory.

facilities that have the financial means to build and operate the associated power supplies and cooling systems, as well as meet the other operating and maintenance costs of these magnets, which are substantial.

To control energy costs, resistive DC magnets are energized only when experiments are conducted with them and thus are routinely cycled between their energized and unenergized states. This cycling leads to fatigue, limiting their operating lifetimes, which are typically 1,000 to 2,000 hours. It is reasonable to anticipate that resistive DC magnets that produce fields significantly higher than 33 T will eventually be built. Before this happens, however, new materials will have to be developed that combine high mechanical strength and low resistivity, and new structural designs may be required.

Outlook for Resistive DC Magnets

As described above, the difficulty of supplying sufficient power and cooling to a resistive electromagnet severely limits the development of higher-field resistive DC magnets. The physical strengths of the materials used within the magnet are also near their limits. Nevertheless, resistive DC magnets will continue to be important because record steady-state fields can be achieved by pairing resistive magnets with superconducting magnets. The Tallahassee site of the NHMFL is where most of the work on resistive magents is now being done in the United States. The objective of this work is to improve the field strength of the resistive outsert of NHMFL's 45-T hybrid magnet (see the section "Pulsed Magnets").

PULSED MAGNETS

Pulsed magnets have long played an important role in high-field science, primarily because they can generate much stronger magnetic fields than DC magnets. However, they do so for only short periods of time (10 to 100 ms), and they tend to have very small bores (10-20 mm). Pulsed magnets are resistive magnets that take advantage of the capacity of materials and power systems to withstand extreme conditions transiently. Figure 3.5 is a plot of the peak magnetic field produced by the pulsed-field magnets now operating around the world versus the duration of the field pulse generated (see Appendix B for more information). By providing access to fields far higher than those accessible using steady-state magnets, pulsed magnets offer opportunities for new discoveries that can take research in new directions; detailed follow-up with steady-state fields is often required, however.

The 60-T long-pulse magnet at NHMFL at Los Alamos National Laboratory (LANL) is a good example of a device of this type (Figure 3.6 shows its (massive) power supply). It produced field pulses having a strength of 60 T for 100 ms in a

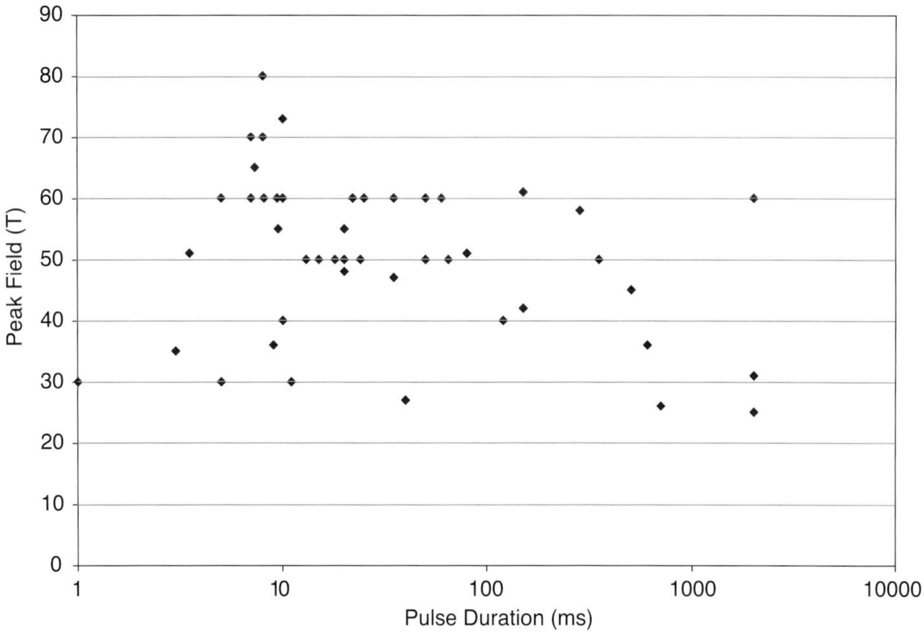

FIGURE 3.5 The field strength and pulse duration times of representative nondestructive pulsed magnets from facilities around the world (see Appendix B).

32-mm bore before failing prematurely in 2000. Pulsed magnets are often precooled to cryogenic temperatures using liquid nitrogen and then energized with a pulse of electric current obtained from a large capacitor bank. Pulse length is limited by the heat capacity of the magnet and by the adiabatic temperature rise of its windings.

The generation of high pulsed fields requires advanced materials for conductors and careful attention to stress management. The ideal conductor would combine high mechanical strength with high electrical conductivity and high heat capacity, properties that are unhappily in conflict, so compromises must be made. Whatever the conductor used, magnets are usually built up in nested layers, each designed to withstand its own radial force by hoop tension and each often reinforced structurally by shells, bands, or fibers of nonconducting or low conductivity material. Design parameters for each layer, such as conductor size, material, current density, and reinforcement, are all varied to optimize magnet performance.

Pulsed magnets that operate in the 50- to 60-T range are relatively common, but only a few facilities have, or plan to build, magnets that produce fields that are much higher, because 60 T is the field strength at which the mechanical strength of the materials used becomes limiting. Nevertheless, pulsed magnets have been built that achieve 70 T, but their pulse lifetimes are extremely short. As this report is being written, two institutions (NHMFL at Los Alamos and the High Field Laboratory at Dresden, Germany) are developing 100-T, multishot pulsed magnets. One approach is to build a magnet consisting of a large outer coil that produces a long pulse with a peak field about 55 T and a small inner coil that generates a much shorter 45-T pulse in synchrony.

Fields substantially in excess of 100 T can be produced by pulsed methods. They are obtained by discharging large amounts of electric current through single-turn coils. The resulting magnetic field pulse ends when the coil vaporizes. Because the plasma so created is driven outwards by magnetic forces, the experimental sample may survive.

FIGURE 3.6 The 1,430-MVA electric generator used by the NHMFL-LANL for pulsing very high magnetic field resistive magnets. Photo by K.N. Roark, Los Alamos National Laboratory.

Not surprisingly, multishot pulsed magnets suffer from cycling-induced mechanical and electrical fatigue. To alleviate these problems, some designs allow for limited plastic deformation of the windings. Electrical insulating materials must withstand cycles of high voltages and mechanical stress. Discontinuities in windings, structures, and insulation systems create mechanical and electrical stress concentrations where failure can occur. They are usually found at transition regions in windings, joints between subcoils, and at coil terminals. Given the magnitude of the energy stored in these magnets when they are fully energized, their end-of-life failure modes are often sudden and highly destructive (see Figure 3.7). Protection of users and surrounding equipment is an important consideration for this class of magnets.

Advantages and Disadvantages

Pulsed magnets are in widespread use today, especially outside the United States. The rapid development of instrumentation in recent years, notably in high-speed electronics and optical detector arrays, has greatly increased the range of work that can be undertaken in pulsed fields. Some topics in semiconductor physics are particularly well suited to such measurements since, due to the low concentrations of free carriers, there are no particularly severe problems associated with sample heating. Additionally, advances in the speed and miniaturization of electronics have increased the range of experiments that can be usefully performed using pulsed-field magnets. However, in general, experiments at the limits of sensitivity and high resolution can only be performed using steady or slowly varying magnetic fields. As developments in technology make measurement instrumentation smaller and faster, though, pulsed fields will continue to offer a valuable alternative. In fact, 16 of the 28 foreign high-field facilities surveyed in Table B.1 in Appendix B provide *only* pulsed-field magnets for their users, and only 2 of them employ DC magnets exclusively. There are many reasons for this bias.

Advantages

Pulsed-field magnets are cheaper to build and operate than high-field DC-magnets. For fields in the 30-T range, where there is overlap with what can be obtained from resistive DC magnets, pulsed magnets are generally much cheaper to construct and operate largely because they require much less infrastructure (power supplies and cooling facilities).[1] For facilities that operate many pulsed

[1]The experience of C.C. Agosta and his students at Clark University provides an example, albeit extreme. They recently built a pulsed magnet that delivers a peak field of 50 T with a rise time of 12 ms. The complete system, including the power supply, cost less than $150,000.

FIGURE 3.7 The fatigue stresses in high-field pulsed magnets often lead to dramatic failures. Images courtesy of National High Magnetic Field Laboratory.

magnets, a further economic advantage arises from the fact that a single power supply can often service several magnets in parallel.

Improvements in instrumentation have significantly increased the effectiveness with which pulsed fields can be used. Because of progress in electronics, instrumentation available commercially today can acquire and store in memory far more data within the span of a single magnetic field pulse than was previously possible. Significant numbers of observations can now be made and recorded within a few milliseconds, and this capability has unquestionably improved the scientific utility of pulsed magnets. In some cases, the dependence of some property of a sample on magnetic field strength, from zero field to the maximum field of a pulse, can be determined today in a single shot. Electrical transport and optical studies are easily done using pulsed-field magnets, and recently an NMR experiment was reported that used a pulsed field of 33 T with a rise time of 10 ms.[2]

Pulsed-field magnets can generate much higher fields than DC magnets. At the present time, the only magnets available that can deliver fields greater than ~45 T are pulsed magnets, and most of the science done with these devices is done using magnets that deliver relatively long pulses. The MegaGauss Laboratory at the Institute for Solid State Physics at Kashiwa, Japan, is an impressive exception in this regard. It offers users routine access to destructive, short-pulse magnets that deliver field strengths up to 622 T. There is important science to be done at these very high fields. For example, there is reason to believe that superconducting DC magnets incorporating high-temperature superconducting (HTS) wire will deliver fields higher than those built using ordinary, low-temperature superconducting wire. Before any of these HTS materials can be used this way, their critical fields and currents (H_c and J_c) will have to be determined experimentally at all temperatures below their critical temperatures (T_c). Pulsed field magnets will have to be used for much of this research because many of the materials under consideration—for example, (1) oxide HTS materials, such as YBCO, and (2) the more recently discovered MgB_2—have critical fields far higher than those generated by any DC magnet. In general, the access to higher fields offered by pulsed magnets can lead to the discovery of new phenomena.

Disadvantages

All the above notwithstanding, pulsed magnets are not the solution to all problems. Pulsed-field magnets cannot be used to measure field-dependent

[2]J. Haase, D. Eckert, H. Siegel, H. Eschrig, K.H. Muller, and F. Steglich, *Concepts in Magnetic Resonance* Part B, 19B(1), 9-13 (2003).

properties of materials that do not equilibrate in times less than the pulse width. Examples of the kinds of measurements for which pulsed magnets are not optimal are measurements of specific heats and thermal conductivities, although new techniques have recently been developed for use in 60-T long-pulse magnets. Quantum Hall effect measurements have also proved to be very difficult in pulsed fields, and DC magnets are essential for most NMR and MRI applications.

Pulsed-field magnets are awkward to use for measurements that require signal averaging. The repetition rate of pulsed-field magnets is very low, often less 0.1 min^{-1}. Thus, if signal averaging is required for some measurement, data accumulation on pulsed magnets is likely to be very time consuming. For all such measurements, DC magnets will be preferred, provided they can deliver the field strength required.

The Potential for Expanding the Use of Pulsed Magnets

Discussions with NHMFL users, who mainly come from the materials science community, revealed their concern about access to the DC magnets at NHMFL, which reflects both the high user demand and the restricted operating schedule of the laboratory. The way DC magnets are used at NHMFL ought to be analyzed to find out if a significant fraction of the experiments could have been carried out using pulsed magnets. If that fraction is large, an obvious way to alleviate the current bottleneck would be to offer users improved access to appropriately equipped pulsed magnets. A robust pulsed magnet with a bore of 25 mm, a peak field of 45 T (equal to today's highest DC field), a pulse rise time of 15 ms, and a fall time of 100 ms would allow exploring the highest magnetic fields in ways that could take research in new directions. Additionally, advances in high-speed electronics, instrumentation, and miniaturization also offer the potential to allow greater experimental access to higher fields. Such a pulsed magnet could probably be built for considerably less than $500,000; the alternative—namely, the construction of a new 45-T DC hybrid magnet—would be vastly more expensive. Another approach to improving user access to high fields would be to increase the operating hours for the existing DC magnets.

Outlook for Pulsed Magnets

Despite their limitations, pulsed magnets produce the highest magnetic fields and are an important part of the nation's high-field portfolio. The scientific motivation for carrying out experiments at the highest available fields is compelling, and the possibility of observing qualitatively new science in this regime is very real. A 100-T, long-pulse magnet would have a good chance of achieving this goal. By "long pulse" is meant a field pulse that is reasonably stable for at least 100 ms. A highly focused materials development program, coupled with an intensive program of high-field magnet design, analysis, and engineering, will be required to make a substantial leap forward in the magnetic fields produced by pulsed methods. A large investment in the power supply and conditioning equipment will also be needed since power demand increases as the square of the magnetic field. In this connection it should be pointed out that it is no accident that LANL is the home of the nation's foremost pulsed magnet program. The LANL program (see Figure 3.6) takes advantage of some remarkable energy generation and storage equipment that became available at Los Alamos only because of the cancellation of an unrelated fusion experiment. A decision to expand the capabilities of the Los Alamos power system would have to be based on a careful evaluation of existing opportunities as well as consideration of greenfield ventures.

SUPERCONDUCTING MAGNETS

The first superconducting magnets were built about 40 years ago, and great progress has been made since then. The many important successes achieved in the past 2 years make one optimistic about the prospects for constructing superconducting magnets in the next decade that are much more powerful than any in existence today.

High on the list of relevant successes is the host of advances in high-field applications of Nb_3Sn. For example, the model central solenoid coil for the International Thermonuclear Experimental Reactor (ITER) fusion experiment, which is by far the largest fast-ramping superconducting magnet ever made, has achieved 13 T with an 8-s ramp-up to full current. It has a 46,000-A, force flow, fully force supported conductor made of 720 superconducting Nb_3Sn and 320 copper strands. When the international decision is made to proceed with ITER construction, the annual production of Nb_3Sn wire will have to increase severalfold, a development that will probably reduce the cost of this critical material for all users. In addition, with the support of the particle accelerator community, Nb_3Sn conductors were developed recently that operate at a current density of 3×10^5 A/cm^2 at 12 T, which is more than double what was possible 5 years earlier. A dipole magnet has been constructed from this material at Lawrence Berkeley National Laboratory (LBNL) that operates at 16 T, a field scarcely thought possible a few years ago. The Large Hadron Collider (LHC), which is scheduled to come online in 2007, uses Nb-Ti magnets that operate at about 9 T in superfluid helium. The possibility that the center of mass collision energy of the LHC might be doubled by replacing those magnets with Nb_3Sn magnets of the LBNL type has stimulated a new European program in Nb_3Sn conductor technology. Finally, the latest generation of high-field NMR magnets operate in a persistent mode at 21.4 T (900 MHz) at about 1.8 K. Better Nb_3Sn conductors that incorporate insights gained from the high-energy physics (HEP) and fusion programs should make it possible to build 1-GHz NMR systems.

An advance of a different sort was made recently by an Oxford Instruments/ NHMFL team, which built a 25-T magnet containing both Nb_3Sn, and Bi-2212 coils. Bi-2212 is a high-temperature superconductor, and 25 T is a new record for a superconducting magnet. The design used should make it possible to build magnets that operate at 30 T at least. The HTS technology used in this magnet is being fostered by the Department of Energy's commitment in 2003-2004 to a major expansion of its industrial superconducting partnership initiative. This initiative should lead to the construction of three underground power cables based on Bi-2223 that have power-handling capabilities exceeding 100 MVA with voltages up to 138 kV. At the same time, a 100-MVA superconducting generator that uses Bi-2223 is being built by GE. Long-term hopes for the widespread application of superconductors in utility applications depend on development of a new generation of low-cost HTS superconductors using novel designs that layer $YBa_2Cu_3O_{7-x}$ thinly on a metal alloy substrate, yielding dramatic improvements in critical current density. The first industrial production of lengths of this material using scalable, continuous processes took place in 2003.

Finally we note the demonstration in 2003 of critical field behavior in a new superconductor, MgB_2, that has upper critical fields exceeding those of Nb_3Sn at all fields and temperatures. MgB_2 suffers neither from the grain boundary, weak-link problems of most HTS materials, which dictate that they be made in strongly textured, often tapelike geometries, nor from the large flux creep that makes most HTS materials unsuitable for operation in the persistent mode, which applications like solution NMR require. Many groups have already prepared prototype wires of this substance. MgB_2 offers some exciting near-term opportunities.

Superconductors Used for Magnet Construction

Several thousand materials are known to be superconducting under appropriate conditions, but virtually all superconducting magnets have been made from just three of them: (1) the body-centered, cubic solid solution alloy Nb-Ti (47 wt% Ti), with a T_c of 9 K, (2) the cubic A15-structure, intermetallic compound Nb_3Sn, with a T_c of 18 K, and (3) the orthorhombic trilayer cuprate $(Bi,Pb)_2Sr_2Ca_2Cu_3O_{10-x}$ (Bi-2223), with a T_c of 110 K. Occasional use is made of Nb_3Al ($T_c \approx 17$ K; see subsection "Materials Occasionally Used" for more discussion) and the bilayer cuprate version of Bi-2223, $Bi_2Sr_2CaCu_2O_{8-x}$ (Bi-2212) ($T_c \approx 90$ K), and both are available commercially on special order. Significant effort has been mounted worldwide since the mid-1990s to make conductors of the less anisotropic cuprate $YBa_2Cu_3O_{7-\delta}$ (YBCO), with a T_c of 92 K, in the form of a quasi single-crystal, multilayer thin film known as a coated conductor. The hexagonal AlB_2-structure compound MgB_2, with a T_c of 39 K, discovered to be superconducting only in early 2001, is now attracting serious attention as a competitor for both the lower-T_c Nb-based conductors and the higher-T_c Bi and Y cuprate conductors.

As will become clear below, the performance of electrical conductors made from superconducting materials is highly sensitive to fabrication details. The performance of kilometer lengths of conductor may fall far below expectations based on measurements made on laboratory-size samples unless scrupulous attention is paid to every detail of the manufacturing process. One of the reasons for optimism about the future of superconducting magnet technology is that at this point, relatively little is known about how practical conductors made of high-temperature superconductors or MgB_2 can be fully optimized, as is discussed below. Substantial improvements in performance are likely to occur as the community works its way along the learning curve.

Economic Considerations

From the point of view of magnet construction, the critical temperature of a superconducting material is less important than the feasibility of fabricating multikilometer lengths of conductor, at reasonable cost, in a form suitable for magnet construction and with critical current densities (J_c) of more than 10^5 A/cm^2 over a wide range of fields and temperatures. The cost-performance metric is dollars per kiloampere-meter of conductor. The J_c for all such substances is strongly field- and temperature-dependent because its magnitude is determined by the volumetric pinning interaction force F_p ($J_c \times B$, where B is the magnetic field), which results from the interaction of quantized superconducting-state flux vortices and the nanostructure of a superconductor. In principle, a homogeneous superconductor with uniform microstructure does not support long-range, bulk supercurrents, but if it is treated to give it defects where superconductivity is locally depressed or destroyed at scales on the order of the superconducting coherence length ξ, J_c can reach 10^6 to 10^7 A/cm^2. Because ξ lies within the range 0.5 to 5 nm for all useful conductors, even atomic-scale defects can produce the pinning interactions required. Superconducting conductors useful for magnet construction are thus fabricated with very high defect densities ($>10^{16}$ cm^{-2}) and defect dimensions of ξ or less, which makes practical superconductors genuine nanostructured materials.

Additional Conductor Requirements: H_{irr}

In addition to needing high J_c and reasonable costs of fabrication, the superconductors used in high-field magnets must have a high irreversibility field, H_{irr}, since J_c becomes zero at H_{irr}. The definition of H_{irr} is fundamentally different for anisotropic high-temperature superconductors and isotropic, low-T_c superconductors (LTS). The classical LTS materials, Nb-Ti and Nb_3Sn, are both cubic and isotropic, and H_{irr} is defined by the phase transition between the superconducting and the normal states of the material, which occurs at the upper critical field, H_{c2}. Smearing of H_{c2} due to the inhomogeneities that must be introduced to give the materials a high J_c makes H_{irr} ~0.9 H_{c2}. The large structural and electronic anisotropy of the HTS materials leads to H_{c2} anisotropies on the order of 100 for Bi-2212 and Bi-2223, H_{c2} and J_c being much smaller for fields perpendicular to the CuO_2 planes than for fields parallel to them. Large anisotropies are a direct consequence of the electronic structure of the cuprates, whose parent state is insulating. Superconductivity is strongest within the CuO_2 planes and weakest in the charge reservoir layers that dope holes into the CuO_2 layers and make them metallic. The charge reservoir layer in the Bi compounds is a double Bi-O layer, which is itself only poorly conducting. These cuprates can therefore be regarded as multilayers that are composed of dirty, normal metal layers (the Bi-O and adjacent Sr-O layers) alternating with superconducting block layers of CuO_2 and their separating Ca-O layer, which together form a repeating superconducting/normal/superconducting stack. In the lower H_{c2} orientation of fields, which is perpendicular to the planes, vortices within the superconducting state are greatly weakened whenever they pass through Bi-O layers. This weakening, coupled to strong thermal activation at higher temperatures, results in the destruction of bulk currents at fields of H_{irr} much lower than H_{c2}.

The reason there is so much interest in $YBa_2Cu_3O_{7-\delta}$ today is that its charge reservoir layer is metallic, making the anisotropy of its H_{c2} much smaller than for any other HTS cuprate, which in turn strongly raises H_{irr}. The highest T_c cuprate is $HgBa_2Ca_2Cu_3O_{10-x}$, with a T_c of 132 K, but it is even more anisotropic than the Bi compounds just mentioned and for this and other reasons is not being seriously pursued for conductor applications. The accessible superconducting state space of finite J_c for actual or potential conductor materials is indicated in Figure 3.8. Interestingly, MgB_2 appears to exhibit BCS-related superconductivity rather than HTS-related behavior despite its relatively high T_c, just below 40 K. Even though MgB_2 is variably anisotropic, depending on its alloying state, its H_{irr} is about 0.9 H_{c2}.

A useful superconducting conductor must also be electromagnetically stable. Because superconductors with high J_c can shield strong fields, they are thermodynamically unstable when carrying bulk transport currents and thus capable of

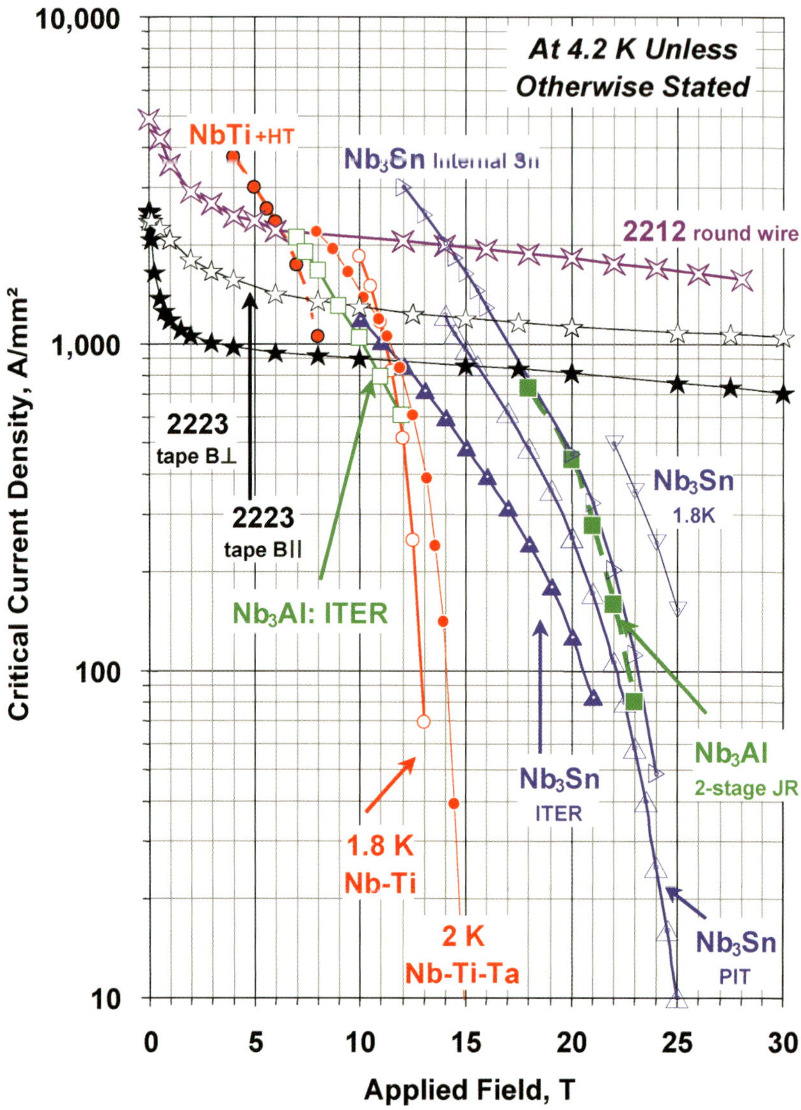

FIGURE 3.8 The field dependence of critical current densities for candidate conductor materials suitable for high-field magnet construction. The data shown were taken using conductors fabricated from the materials indicated. These data are obtained from the University of Wisconsin's Applied Superconductivity Center; the center maintains a database of critical current density measurements and provides free online access at http://www.asc.wisc.edu/plot/plot.htm. Figure courtesy of Peter J. Lee, University of Wisconsin at Madison.

flux-jumping to the more stable state of lower J_c. To prevent this, they must be subdivided so that the stored energy of the shielded field, which is proportional to $J_c d$, where d is the transverse dimension of the superconductor, is small enough not to quench the superconducting state. A high-conductivity matrix of Cu in Nb-base superconductors or Ag in Bi-base conductors provides the stability required both during superconducting operation and during the occasional destructions of the superconducting state that occur in a magnet quench. Thus, a useful superconducting conductor is inherently a composite of superconductor and normal metal in which superconducting filaments are embedded in sufficient normal metal to ensure electromagnetically stable operation. Various ways in which this requirement can be met are shown in Figure 3.9.

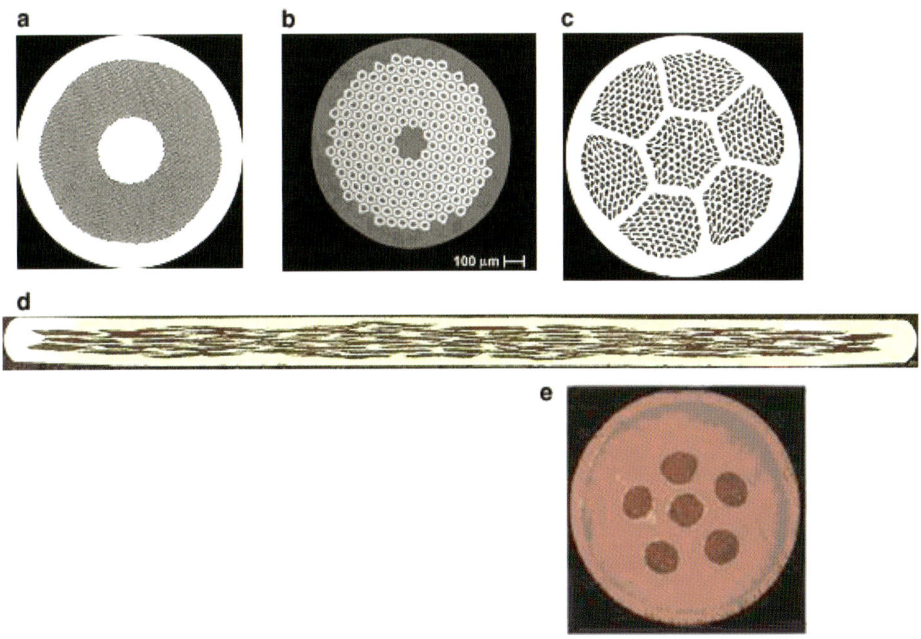

FIGURE 3.9 Representative multifilamentary conductors made from (a) Nb47wt%Ti, (b) Nb_3Sn, (c) Bi-2212, (d) Bi-2223, and (e) MgB_2. The matrices for the conductors are high-purity copper for (a), (b), and the outer sheath of (e) and pure silver for (c) and (d). The filaments of MgB_2 are surrounded by 316 stainless steel in (e). Conductors were manufactured by Oxford Instruments Superconducting Technology (a) and (c), ShapeMetal Innovation (b), American Superconductor Corporation (d), and Hitachi Cable in collaboration with the National Institute for Materials Science (e).

Critical Current Densities

A final important parameter for characterizing the performance of superconducting materials in high-field magnets is the critical current density J_c, the density of current flow that extinguishes superconductivity in a particular sample of a superconductor. Although the critical current density of any sample is limited by the intrinsic properties of the superconducting material it contains, it is highly dependent on extrinsic factors, such as the microstructure of the sample being considered and the way the superconducting material it contains was processed. In the limit of zero applied field and zero temperature, all superconductors useful for magnets have J_c values exceeding 10^6 A/cm^2. For magnets with meter-size bores (e.g., fusion or particle-detector magnets), superconducting wires that have J_c values lying at 10^4 A/cm^2 or more suffice, but for small magnets (e.g., solenoids or beam-steering magnets), where apertures are a few centimeters or so, J_c values of 10^5 A/cm^2 or more are required. The magnitude of J_c as a function of the magnetic field H—$J_c(H)$—is equal to the summed interaction of individual vortex pinnings with the defect sites. Broadly speaking, one can distinguish two situations. In one situation, there is full summation of all interactions over all H-T space. This is the case for Nb-Ti, where flux pinning is well understood and the fabrication process is designed to optimize $J_c(H)$. More generally, however, the density of pinning interactions is less than optimum, and the summation of pinning forces is subject to the collective properties of the vortex lattice. A compendium of $J_c(H)$ data, provided in Figure 3.8, shows that field dependence is generally greater in the collective case. High relative values of $J_c(H)$ are more easily designed and obtained with Nb-Ti. The main-ring dipole magnets of the LHC at CERN are good examples of magnets operating in this regime. The recent Nb_3Sn magnets used in 900-MHz NMR spectrometers are exceptional in operating up to about 75 percent of H_{c2}. An advance beyond 1 GHz is unlikely with Nb_3Sn.

Existing Conductor Materials

Nb-Ti

Nb-Ti is a strong, tough, ductile, well-understood material with a T_c of 9 K, an H_{irr}(4.2 K) of about 10.5 T, and an H_{c2}(4.2 K) of about 12 T.[3] It is the workhorse of the superconducting industry, being used for all MRI and accelerator magnets. It is universally used for fields up to about 8 T, at which point Nb_3Sn becomes

[3]Throughout this report, the notation Nb-Ti denotes the alloy, which can have a range of elemental combinations, normally niobium alloyed with about 47 wt% titanium.

more attractive. If Nb-Ti magnets are cooled to between 1.8 and 2.0 K using superfluid helium, they can operate at fields as high as 12 T. The advantages of Nb-Ti are several. It is well understood how to fabricate wires of Nb-Ti that have optimized nanostructures and high J_c. The alloy is strong and tough, making Ni-Ti conductors robust in fabrication and application. Its cost-performance ratio is low, about $1 per kiloampere-meter (/kA-m) at 5 T and 4.2 K. It can be made in many designs, with filament sizes ranging from about 1 to 100 mm surrounded by matrices of pure Cu or, occasionally, pure Al. If necessary, alloyed Cu can be used to enhance transverse resistivity so that the filaments are less easily coupled by high magnet charge rates or AC use. A typical accelerator conductor containing a few thousand 8-μm diameter Nb47Ti filaments embedded in a Cu matrix is shown in Figure 3.9a. MRI magnet conductors are generally simpler because filament diameters are larger, about 50 μm.

Nb_3Sn

Nb_3Sn is an intermetallic A15 compound that exists over a range of compositions: Nb combined with anywhere from 18 to 25 percent by (atomic) weight Sn. In 1961 its use for high-field superconducting magnets was discovered, and it remains one of the two most important materials for such magnets. A great deal of effort was devoted to its development as a conductor for magnets in the 1960s and 1970s and, again, in the last 3 or 4 years, when great strides were made, as discussed earlier. This developmental work, which has been driven by the needs of three communities—NMR, fusion, and, more recently, high-energy physics—has greatly improved our understanding of Nb_3Sn. Because Nb_3Sn is now very well understood, it is easier than ever to make from it conductors with enhanced J_c and H_{c2}, even though its H_{c2} at liquid-helium temperatures is about 28 T (see Figure 3.10). It is the superconductor of choice for magnets producing fields from 10 T to more than 20 T. It is conceivable that it will be used to build NMR magnets that operate at 23.4 T (1 GHz) but unlikely that it will go much beyond that because of the intrinsic H_{c2} limitation of the material itself.

Nb_3Sn wire can be manufactured in many different forms to suit design requirements, one of which is shown in Figure 3.9b. However, once formed, Nb_3Sn is notoriously brittle. More than that, the critical current densities of Nb_3Sn conductors are a strong function of applied longitudinal strain, an effect that increases with the magnetic field strength. To control these problems, Nb_3Sn coils are usually manufactured by a process called insulate-wind-react. The conductor is produced in a form in which its Nb and Sn components are adjacent but separate. Electrical insulation is wrapped around the conductor while it is in this ductile state and the conductor is wound into the coil form required. The completed coil is then treated

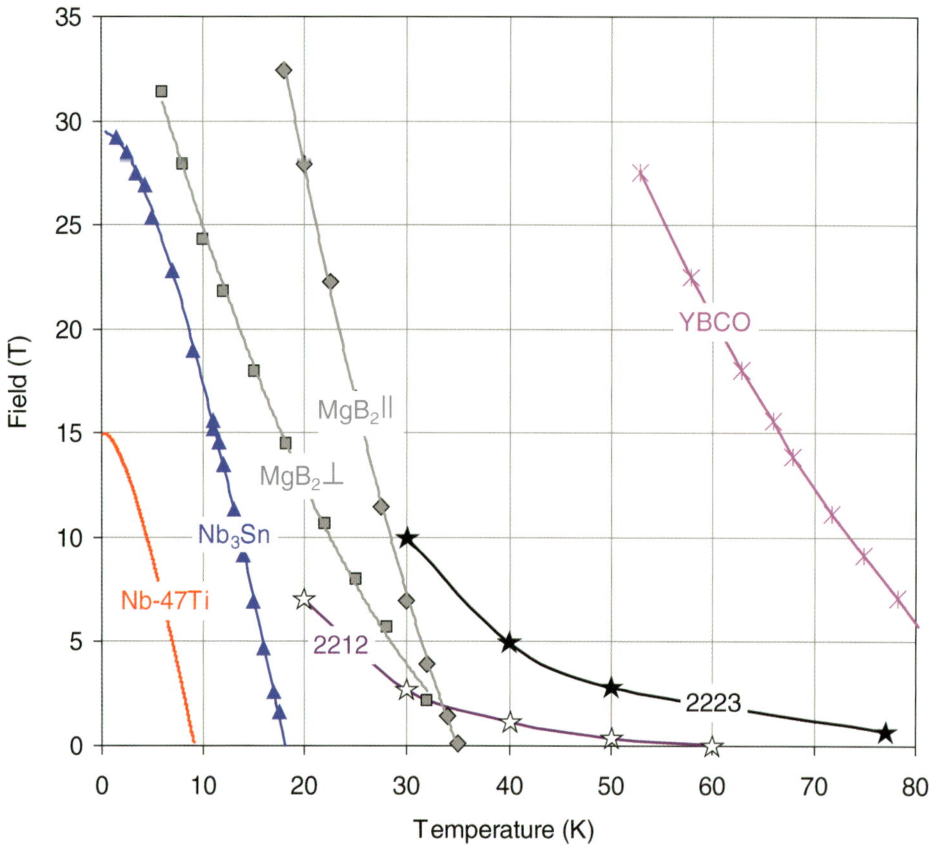

FIGURE 3.10 Upper critical fields (H_{c2}) for Nb47wt%Ti, Nb_3Sn, and MgB_2 and irreversibility fields (H_{irr}) for Bi-2223 and YBCO. Note that the two fields are very close for Nb47wt%Ti, Nb_3Sn, and MgB_2 (H_{irr} is 85-90 percent of H_{c2}) but far apart for all cuprate superconductors. For MgB_2, Bi-2223, and YBCO, the values plotted are the lower values appropriate for fields perpendicular to the strongest superconducting planes, the B planes for MgB_2 and the CuO_2 planes for the cuprates. Values plotted are the highest found for each compound.

at a high temperature, which causes its Nb and Sn components to mix, forming Nb_3Sn. Extreme care must be taken with the coil once it has reached this stage because the wire in it is no longer ductile. Large thermal stresses may develop when such a magnet is cooled to cryogenic temperatures, and when the coil is energized, Lorentz forces create additional stresses. All of these effects must be appropriately anticipated if a useful magnet is to result.

Considerable effort has been expended to find optimum conditions for the heat treatment of Nb_3Sn conductors. Treatment at 650-700°C is compatible with the limitations imposed by using glass-fiber insulation, so Nb_3Sn treated like this is suitable for the construction of small-bore magnets for NMR and laboratory use. This rather low temperature has the additional advantage of producing a fine-grained (~100-nm) Nb_3Sn phase that provides grain boundary vortex pinning, which maintains a higher J_c. However, at temperatures this low, the Sn content of the A15 phase does not equilibrate, and gradients of Sn content occur across filament layers, which are a few microns thick; these differences lead to gradients in the superconducting properties, complicating the optimization of wire properties.

Bi-2223

Bi-2223 conductors of the sort shown in Figure 3.9d are the first and so far the only fully commercial HTS conductor material. The highly aspected shape (~20:1) of the conductor follows that of the filaments of which it is composed and of the crystal structure of $(Bi,Pb)_2Sr_2Ca_2Cu_3O_{10-x}$ itself. One consequence of the parent insulating state of HTS materials is that grain boundaries tend to be weak-linked unless there is a very low level of misorientation. Making the conductor as a highly aspected tape favors a high degree of texture. However, many obstacles to current flow still remain in such conductors, even when they are textured. Supercurrent must thus percolate through a network of such obstacles, reducing the expected current density values by a factor of 5 or so. As Figure 3.8 shows, J_c values for Bi-2223 do not yet compete with those of Nb-based conductors, even at 4.2 K, owing entirely to the reduction of useful cross section produced by barriers to current flow, the most important of which are more angular grain boundaries.

HTS conductor development has largely been motivated by the desire to replace Cu and Fe in utility applications. Since such devices work either in self-fields of about 0.1 T for power cables or 1 to 2 T for transformers, motors, and generators, Bi-2223 can be used at quite high temperatures, though as Figure 3.10 shows, H_{irr} falls rapidly with field, precluding multitesla applications above 30 K or so. Even though the T_c is 110 K, it can only be used in weak fields at 77 K (boiling liquid nitrogen).

Bi-2223 is the workhorse of virtually all HTS utility applications (there are a few uses for Bi-2212), but it suffers from having to compete in cost with Cu and Fe. It is estimated that Bi-2223 conductors would have to cost ~$10/kA-m at 77 K in order to compete with Cu and Fe, but because of its still relatively low J_c and the high (about 60 percent) Ag content of wires made of it, it costs 10 times as much.

While Bi-2223 is the best HTS conductor available today, it is unclear whether it will remain so. Bi-2212 has better connectivity than Bi-2223 for reasons that are not well understood and thus may emerge as a better low-temperature, high-field conductor than Bi-2223. In the long term, both substances may be replaced by biaxially textured YBCO, which has a much more favorable $H_{irr}(T)$ curve (see Figure 3.10), lower anisotropy, and higher J_c (see Figure 3.8) and does not require Ag coating to be an effective conductor.

Materials Occasionally Used

Many other materials have been found that will superconduct under certain conditions and that might be used for magnet construction—for example, ternary alloys such as Nb-Ti-Ta, which is being studied as a possibility for accelerator magnets. Rather than discussing them all here, we will mention only one, the binary compound Nb_3Al.

Nb_3Al has a number of potential advantages over Nb_3Sn, not least that the stoichiometric compound has a slightly higher T_c and an H_{c2} that can reach 40 T. Its greatest shortcoming is that it does not easily form when it contains 25 wt% Al, such that when quenched from the high temperatures at which the stoichiometric composition is stable, it forms a disordered structure in which both T_c and H_{c2} are lower. An important advantage of Nb_3Al is that the strain sensitivity of its J_c, T_c, and H_{c2} parameters is much less than that of Nb_3Sn.

There are two principal approaches to making Nb_3Al conductors. One uses jelly rolls of Nb and Al that are extruded and drawn into wire before being reacted at 700 to 900°C to form the A15 phase. The conductor that results has a T_c of about 16 K and an H_{c2} of 20 to 24 T, which are both significantly lower than for competing Nb_3Sn conductors. Because the properties of Nb_3Al are little affected by strain, conductors made of it have been used in prototype coils for large fusion magnets. The second approach, which was pioneered in Japan, is to rapidly heat Nb-Al precursors to about 2000°C and then quench them in a liquid Ga bath. This maintains a ductile Nb-Al solid solution phase that can be then selectively crystallized into an off-stoichiometric but still high J_c and high H_{c2} phase at lower temperatures. Some examples of the high J_c values that can be obtained by such a process are shown in Figure 3.8. It is, in fact, possible to find a range of magnetic field above ~20 T where Nb_3Al is superior to Nb_3Sn, and Japanese researchers are pursuing the use of such conductors in high-field NMR magnets. A variety of problems remain to be resolved. Among them are the viability and costs of such a complex process and finding a means of applying sufficient stabilizing normal metal—e.g., Cu or Au—to the conductor.

Emerging Superconducting Materials

While the normal and superconducting states of HTS materials remain incompletely understood, work on their practical application continues to advance. In fact, in many applications, HTS materials exhibit surprisingly conventional behavior. The property of HTS materials that has presented the greatest barrier to their practical use is the intrinsic weakness of the links at grain boundaries. This weakness makes the supercurrent density at grain boundaries less than within grains, reducing the overall supercurrent in conductors made of polycrystals. This problem can be alleviated by using epitaxial films in which grains are aligned in the plane of the substrate. Indeed, HTS epitaxial films are used in the wireless communications filter systems deployed today, with a few thousand such systems having been sold. For applications requiring a conductor (e.g., magnets, power transmission, energy storage, motors, and generators), tapes and wires of Bi-based HTS conductors are used, and the performance of these conductors has steadily improved. More recently, attention has focused on the development of tapes that consist of a thick film of YBCO deposited on ribbon substrates in which the grains have been aligned by various means so that grain boundary misorientation angles are small. Finally, discoveries about the superconducting properties of MgB_2 have catapulted it to the fore. These useful developments notwithstanding, the industrial application of HTS materials is hampered by other physical issues, one being the need for expensive cooling systems to maintain them in the superconducting state and another being their brittleness.

Bi-2212

$Bi_2Sr_2CaCu_2O_{8-x}$ is a two-CuO_2-layer version of Bi-2223. It shares many of the features of the latter compound but has enough unique advantages so that it is still produced commercially in small quantities. One of its advantages is that it can be made as a round wire with high J_c. It appears that considerable self-alignment of Bi-2212 grains occurs during the partial Bi-powder precursor melt step in the manufacturing sequence, which is required to obtain a high J_c. Indeed, the J_c values of Bi-2212 round wires considerably exceed those of Bi-2223 wires. As Figure 3.8 shows, the field dependence of J_c for this material is very weak up to 30 T at 4.2 K. Indeed the very high field limits of this material, which are only poorly understood, are probably well above 50 T at 4 K. This is a direct consequence of the excellent connectivity of Bi-2212 conductors. A typical round wire conductor is shown in Figure 3.9c.

Some manufacturing problems remain to be worked out. For example, the temperature window for optimum processing is very tight, perhaps as little as 5°C at 890°C. This makes the wind and react process that has proven so successful for Nb_3Sn magnets more difficult to execute for Bi-2212. Also, like all conductor materials except Nb-Ti, Bi-2212 is brittle and, even when sheathed in Ag or Ag alloy, not very strong.

The 25-T superconducting magnet made by an Oxford/NHMFL team in 2003, which was mentioned earlier, demonstrates the application of Bi-2212 to small-bore solenoids. This technology is likely to make it possible to construct superconducting magnets that operate at 30 T. However, the large electronic anisotropy and strong flux creep of Bi-2212 make it unlikely to be a useful, persistent-mode material and may limit its use to driven magnets or to solid-state NMR. Builders of HEP magnets are also interested in Bi-2212 because the round shape of the wire makes it possible to cable 20-30 such wires into multikiloampere conductors suitable for accelerator dipole or quadrupole magnets.

YBCO-Coated Conductors

One key incentive for the study of YBCO-coated conductors is their exceptionally high J_c value, as shown in Figure 3.8. These values reflect the fact that biaxially textured conductors can be made of YBCO, in which there is almost no grain-boundary obstruction of current. The inherently low electronic anisotropy of YBCO (from 5 to 7) produces excellent flux pinning, giving it strong advantages in superconductor current density over any other HTS compound, including the Bi-based conductors. Prototypes of tape conductors made of this material are now in advanced development by several companies worldwide, suggesting that effective commercialization is imminent (see Figure 3.11).

Complicating any evaluation of the prospects for YBCO-coated conductors is the fact that coated conductors are fundamentally different from any other conductors used for magnet construction (compare Figures 3.8 and 3.10). In contrast to all earlier conductors, which are made by conventional metalworking processes such as extrusion and wire drawing, a coated conductor is made by sequential deposition of a multilayer oxide buffer, which is interposed between the Ni-alloy substrate and the YBCO. Purpose-built production lines of a new kind will therefore be needed. An additional issue with coated conductors is that only about 1 percent of the cross section of such wires is superconducting. Thus, overall conductor current densities are heavily diluted compared with competing conductors, where the superconductor fraction is typically 25 to 50 percent. However, the development of coated conductors is proceeding rapidly, such that prototype lengths have overall current densities that are competitive with the Bi-2223 conductors that they are designed to replace, even with such small fill factors. It is clearly encouraging that a very high current density, very high H_{c2} conductor is possible

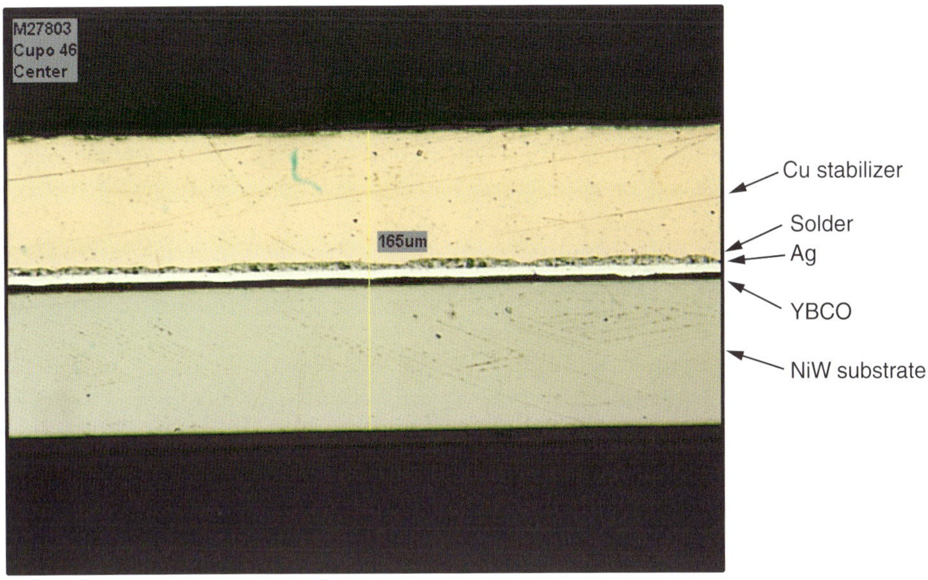

FIGURE 3.11 Illustrative longitudinal cross-sectional architecture of a 165-μm thick, neutral-axis, YBCO-coated conductor showing the copper stabilizer, the solder joining the Cu to the YBCO, and the Ni5W-textured template (a nickel-tungsten alloy with tungsten at 0.5 percent). The actual YBCO superconductor layer is about 1 μm thick and thus represents less than 1 percent of the cross section, in comparison with competing Nb-base, Bi-base, and MgB_2 conductors, in which 25 to 50 percent of the cross section is superconductor. Image courtesy of American Superconductor Corporation.

with YBCO. As lengths longer than the present 10- to 100-m prototype lengths appear, it will become possible to make test coils that can be used to assess the true promise of YBCO-coated conductors for high-field magnets.

Magnesium Diboride

The most recent entry into this field is MgB_2, an apparently simple binary compound that was discovered to be superconducting only in 2001. Its potential application space is indicated in Figure 3.10, which shows the transition temperature of MgB_2 to be about twice that of Nb_3Sn (35-39 K vs. 16-18 K). It can be alloyed so as to have a critical field superior to that of Nb_3Sn at all temperatures. The superconducting characteristics of MgB_2 are interesting not only because of its relatively high transition temperature but also because it has more than one superconducting energy gap, a state of affairs anticipated in BCS theory but never before seen experimentally.[4] MgB_2 is a metal with a layered structure in which the boron atoms form hexagonal layers and the magnesium atoms are located between the boron layers, above and below the centers of the hexagons. The unusual properties of the material are thought to result from strong interactions between phonons and electron orbitals in the boron layer. But unlike previously known good superconductors, MgB_2 does not have a high charge-carrier density, only simple *s*- and *p*-shell electrons are involved in its superconductivity, and it has a noncubic crystal structure. Other features make MgB_2 even more remarkable: Only a few of the atomic-vibration modes are involved in its ability to superconduct, and only about half of its charge carriers are strongly affected during superconductivity.

The remarkable properties of MgB_2 open a new window in superconductivity for fundamental as well as applied research. Prospects for the applications of MgB_2 are enhanced by the inherently low cost of Mg and B. Conductors can be made by placing either B + Mg or MgB_2 powders into metal tubes that are then drawn out into wires such as that shown in Figure 3.9e using conventional metal fabrication technology. Wires up to 1 km long have already been made by two or three companies in Japan, and this success has stimulated the development of lower-field, higher-temperature magnets for MRI. Furthermore, the density of MgB_2 is comparable to that of aluminum, which may lead to new applications where weight is an important consideration. Present conductors do not yet have the capability indicated by the thin films whose H_{c2} data are shown in Figure 3.10, and the current densities achieved are not yet fully competitive with those of Nb-based conductors. Nevertheless it is clear that with proper alloying and development, MgB_2 could outperform any Nb-based conductor. For these reasons, the HEP magnet community has started to explore options for using MgB_2 in the next generation of accelerator magnets. Equally exciting, the critical temperature of MgB_2, 39 K, would allow electronic circuits based on this material to operate at 20-25 K, achievable by a compact cryocooler, which gives this material a significant advantage over low-temperature superconductors. Compared with the high-temperature superconductors, MgB_2 is simpler, cheaper, and more stable over time.

Superconducting Magnet Design

Superconducting magnet technology has advanced to the point where 20-T laboratory magnets are not uncommon, and 900-MHz NMR magnets are available commercially. As noted earlier, a 920-MHz NMR magnet (21.6 T) having a 54-mm bore has been put into operation at the National Institute for Materials

[4]H.J. Choi, D. Roundy, H. Sun, M. Cohen, and S.G. Louie, The origin of the anomalous superconducting properties of MgB_2, *Nature* 418, 758-760 (2002).

Science (NIMS), Tsukuba (Japan), and a 900-MHz, 63-mm bore magnet manufactured by Oxford Instruments is operating at Pacific Northwest National Laboratories (PNNL). The NHMFL in Tallahassee recently brought to full field its 104-mm bore, 21.15-T magnet, which is the largest bore 21.15-T magnet with persistent field in the world. The next big step is clearly going to be the development of 1-GHz (23.5 T) NMR magnets, but their size, complexity, and cost will be daunting.

High-field superconducting magnets present many of the same engineering problems as resistive and pulsed magnets—namely, very high stored energy and stresses—but they also have problems all their own. The superconductivity of superconducting materials is limited by the strength of the magnetic field they experience in a manner that depends on their operating temperatures, and all the superconducting materials now in use, except Nb-Ti, are mechanically weak, or intolerant of strain. In addition, today's superconducting magnets require liquid-helium cooling systems that must be integrated into the coil without compromising its electrical or structural integrity. Even more challenging is the requirement that the energy in the field of an energized superconducting magnet be safely and passively dissipated in the event the coil quenches (i.e., ceases to be superconducting). The energy, which can amount to many tens of megajoules, must be dissipated within the magnet or to external energy-dump circuits. Failure during quench can occur due to overvoltage because of a too-rapid collapse of the magnetic field or, at the opposite extreme, because the magnetic energy dissipates in the generally small normal-state regions of the coil. To design superconducting magnets to withstand these perils is a complex but essential undertaking.

Some of the same engineering solutions to the structural problems of resistive magnets have been adopted for superconducting magnets. For example, magnets are often divided into radially independent subcoils to reduce the transmission of radial forces and to limit hoop stress. This strategy can also result in a more efficient use of materials because the coils can be "graded." More expensive superconducting materials such as Nb_3Sn can be used in the inner coils, while less expensive, more ductile conductors like Nb-Ti can be used in the outer coils. In addition, conductor size and number of turns can be varied in each subcoil to optimize the winding current density, and mechanical reinforcing structures can be incorporated within each subcoil as needed. These techniques are generally used for high-field NMR magnets because of their persistent-mode operation. Magnets intended to produce more transient fields often use internal heaters or external shunt resistors.

Mechanical strain is an extremely important issue for superconductors, as already noted, and electrothermal stability is another critical factor in the design and operation of superconducting magnets. All the solid components of a superconducting coil have extremely small heat capacities at 4 K and below, which means that even a slight disturbance, such as the microheating generated by a local stick-slip friction event, can cause a small portion of the conductor to go normal, which generates resistive heating and may cause the entire coil to quench. This behavior requires a mechanical design different from that for a pulsed magnet, where some plastic deformation of the conductor and coil can be allowed.

Outlook for Superconducting Magnets

Experience shows that Nb_3Sn reaches its practical limit of usable current density in the 21- to 23-T range, a limit that many current superconducting magnets have now reached. Nb_3Sn technology might be further developed to the point that 1-GHz NMR magnets can be made using it, but it is unlikely that it will go much higher than that because of the intrinsic limitation on H_{c2}. A new superconducting material will probably be needed. An A15 compound such as Nb_3Al might be suitable; this material is being developed at a very modest pace and has yet to be produced in large quantities. However, it is widely believed that some form of one of the newer HTS superconductors or MgB_2 will emerge as the conductor of choice. Although these materials are widely recognized for their high critical temperatures, the more relevant measure of their usefulness is their very high upper critical fields. Like Nb_3Sn, however, these materials are all weak structurally. Stress limitation will therefore be an engineering issue, although the strain tolerance of these materials is high enough to permit winding at reasonable bending diameters.

While the availability of new and improved superconducting materials offers hope that superconducting magnets can be built that deliver fields significantly greater than those available today, the challenges still to be met should not be underestimated. The intrinsic coupling of high magnetic field and high stored energy, leading to high stresses, will force magnet sizes to grow unless the yield strength and modulus of elasticity of materials used for their construction can be increased. Increases in magnet size reduce performance because the farther conductors are from the center of a magnet's bore, the less they contribute to the magnetic field in the bore.

HYBRID MAGNETS

The highest DC magnetic fields available today are produced by hybrid magnets, which consist of an inner, water-cooled, DC resistive magnet surrounded by a superconducting outer magnet (see Figure 3.12). The 45-T (32-mm bore) magnet at the NHMFL, which is the most powerful DC magnet in the world, obtains 31 T from its resistive magnet and 14 T from its superconducting magnet. The outer diameter of the resistive magnet of course determines the bore size of the super-

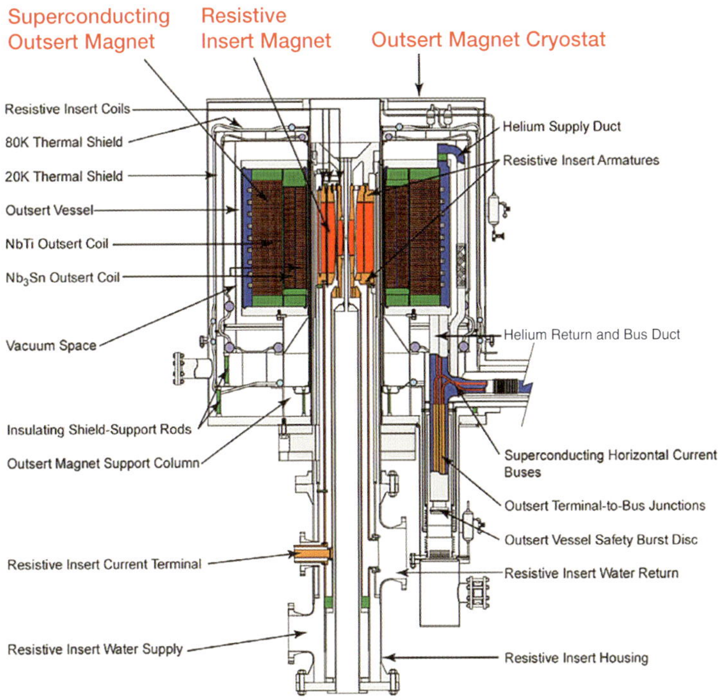

FIGURE 3.12 View of the NHMFL 45-T hybrid magnet. Images courtesy of National High Magnetic Field Laboratory.

conducting magnet. Although the 14-T field produced by the superconducting coil of this magnet is significantly less than the peak field of the NIMS 920-MHz magnet mentioned earlier, the energy stored in its field is three times greater (115 MJ) because its bore is so much larger. The apparatus to support and sustain a hybrid magnet is also much larger; the Florida 45-T hybrid is more than 10 m tall.

Design Challenges

Unusual engineering is needed to solve the stress and cooling problems of a hybrid magnet because there is so much energy stored in its field. For these magnets, field homogeneity requirements are usually relatively modest. The superconductor in the 45-T hybrid is a cable-in-conduit conductor that consists of a multistrand, multistage superconducting cable encapsulated in a structural steel conduit (see Figure 3.13). The conductor is insulated and then wound into the coil form required. If Nb_3Sn is the superconductor used, as is the case for two of the three coils in the superconducting portion of the 45-T magnet at Tallahasee, the reaction heat treatment must be performed after the windings are formed. Cooling is accomplished by forcing supercritical helium (or stagnant superfluid helium-2) through the interstices between cables in the conduit, which account for approximately 35 percent of its total volume. The advantages of this design include (1) very high electrothermal stability because of the high heat capacity of liquid helium and large heat transfer surface area to helium in the conduit, (2) integration of structural steel throughout the winding pack, and (3) high voltage integrity of the coil because of the complete coverage of conductor with insulation, which is achieved without compromising the cooling of the magnet.

The use of a multilevel cable allows for conductor currents of up to several kiloamperes and results in a lower terminal voltage when energy is dumped to an external room-temperature resistor if the magnet quenches or is deenergized for other reasons. However, high conductor current also requires the use of vapor-cooled current leads, which increases either helium consumption and/or refrigeration requirements. The NHMFL magnet also takes advantage of conductor grading to increase efficiency by using conductors of three different sizes for the three subcoils.

One of the main problems of hybrid magnets is protecting their superconducting components from the consequences of sudden resistive magnet shutdowns caused by failure of magnet integrity or of the power supply. The resulting rapid decrease in the resistive magnet field can produce AC losses in the superconducting windings large enough to make them quench. In addition, the large mutual coupling between the magnetic fields of the two coils generates a large Lorentz force, the sudden removal of which may cause mechanical displacement of the superconducting magnet, which could have dangerous mechanical or electrical consequences. Such

FIGURE 3.13 Cross section of one of the cable-in-conduit conductors used in the 45-T hybrid magnet at NHMFL. The stainless steel conduit provides structural support and the 35 percent void around the cable provides a helium cooling area. The overall winding pack current density is reduced by these elements because they cannot carry current to contribute to the magnetic field generated. Image courtesy of National High Magnetic Field Laboratory.

unplanned and disastrous quenches have occurred in the past, and newer designs take this possibility explicitly into account.

Outlook for Hybrid Magnets

NHMFL-Tallahassee is proposing the construction of a series-connected hybrid magnet, the resistive and superconducting components of which will be operated electrically in series from the same power supply. The high inductance of the superconducting magnet should reduce the current ripple in the resistive magnet, thereby improving the temporal stability of the magnetic field. A second important advantage of this design is that no overcurrent or unbalanced forces occur during a quench. In addition, the resistive coil will be designed so that it can be shimmed to higher field homogeneity than normal for a resistive magnet. The series-connected hybrid magnet is designed to produce 35 T (40-mm bore) with an inhomogeneity of less than 1 ppm in a 10-mm-diameter spherical volume. Hybrid magnets such as this one are intended to reduce the power consumed to produce the fields they generate rather than maximizing field strength.

COORDINATION OF MAGNET DEVELOPMENT

One striking characteristic of all of the sciences that use high magnetic fields is how constrained they all are by a single body of technological information—namely, magnet technology. This shared interest notwithstanding, each constituency has historically tended to develop the magnets it needs without much reference to the others. The reasons are several and obvious. The different communities have different missions and need magnets that differ correspondingly. In addition, they are supported by different funding agencies, each with its own perspective.

Recently the HEP and fusion communities have begun to more closely coordinate their development of Nb_3Sn wire, an important first step in a healthy direction. Given the intense interest of all magnet user communities in magnet technology and engineering, it would seem to make sense for them to mount a coordinated effort to advance magnet technology. One component of such a coordinated effort might be the construction of high-field instrumentation at NHMFL specifically for the engineering research necessary for the production of high-performance magnets. A second component would certainly be the development of new materials for magnet construction, and a third would be an exploration of other constraints on magnet design and performance. The final ingredient would be a framework to coordinate and integrate communication among the different communities. This framework might start with something as simple as topical conferences and could extend to a management structure for operating a joint program.

It is clear from recent experience that magnets producing fields significantly higher than those available today are going to be either impossible or prohibitively expensive if current technology is used. Rather than supporting an all-out, brute-force effort to prevail using barely adequate technologies, it makes sense to find new approaches that will make it easier (and cheaper) to build the magnets needed. Essential to this enterprise will be the development of both resistive and superconducting materials with improved electrical, magnetic, and mechanical properties. This initiative could be a collaborative enterprise involving both publicly supported researchers and commercial enterprises, but however it is structured, it will certainly require substantial public support.

4

Conclusions and Recommendations

CONCLUSIONS

Current State and Future Prospects

Conclusion. *High magnetic field science and technology are thriving in the United States today, and the prospects are bright for future gains from high-field research.*

High magnetic field science is having an important impact in many disciplines, including medicine, chemistry, and condensed-matter physics. Recent accomplishments include the development of functional magnetic resonance imaging (fMRI), which is revolutionizing neuroscience; optically pumped magnetic resonance techniques, which allow visualization of new quantum phenomena in semiconductors; and ion cyclotron resonance mass spectroscopy, which is becoming an important tool for exploring the chemical composition of complex systems. High-field research has led to the discovery of new states of matter in low-dimensional systems, and it has also provided the first indications of how high-temperature superconductors evolve into unconventional metallic alloys in the extreme quantum limit. Improvements in ancillary instrumentation and the development of new strategies for using high-field magnets have contributed to these advances and should continue to do so. Outstanding work continues to be done in the area of magnet engineering, the discipline on which all these other activities depend. There is every reason to believe that developments as interesting as these will continue in the decades to come, especially if magnets are built that deliver higher fields than those available today. For instance, discoveries made using pulsed-field magnets, which operate at very high fields and are equipped with instruments that take full advantage of advances in electronics, could take research in fruitful new directions that cannot now be anticipated.

U.S. High-Field Efforts in the International Context

> ***Conclusion.*** *The United States is a leader in many areas of high-field science and technology, but further investment will be required to make it competitive in some critical areas.*

There are many indicators of the strength of the U.S. effort. For example, condensed-matter and materials scientists from other parts of the world routinely travel to the National High Magnetic Field Laboratory (NHMFL) to perform experiments that they are unable to do at home, but U.S. scientists seldom travel abroad for that purpose. An important corroborating observation is found in the European Science Foundation's 1998 report *The Scientific Case for a European Laboratory for 100 Tesla Science.*[1] In that report, a panel of experts was charged with investigating and proving the users' scientific case for such a world-class laboratory. The report concluded that the scientific case for a 100-T laboratory was compelling and that, in fact, one of the prime motivations was "to be competitive with laboratories elsewhere, particularly in the United States and Japan." In addition, the superconducting magnets being installed in the Large Hadron Collider (LHC) at CERN, which were built in Europe, as well as those contemplated for the International Thermonuclear Experimental Reactor (ITER), depend on magnet technology developed in the United States, as do the magnets installed in several other user facilities overseas. However, in the area of NMR, which is a major component of high-field science, the United States is competitive but not dominant.[2] About half of the instrumentation used by NMR spectroscopists in the

[1]European Science Foundation, *The Scientific Case for a European Laboratory for 100 Tesla Science,* ESF Studies on Large Research Facilities in Europe, 1998. Available online at http://www.esf.org/publication/109/100T.pdf.

[2]The issues of (and case for) U.S. international leadership in key research areas and among international leaders in important areas of science and technology is best described in a 1993 report from the National Academies, *Science, Technology, and the Federal Government: National Goals for a New Era*, Washington, D.C., National Academy Press, 1993.

United States, and virtually all of the magnets in their NMR spectrometers, were manufactured abroad. Further, many of the most important recent advances in NMR were made overseas, and, in general, European and Japanese companies have been ahead of U.S. companies in commercializing magnet technology.[3] Finally, Europe is far ahead of the United States in equipping its synchrotron light sources and neutron-scattering centers with instruments for studying the x-ray- and neutron-scattering properties of materials in high magnetic fields. Likewise, Japan operates several key facilities that have played large roles in developing technologies for the highest steady-state and pulsed magnetic fields.

Promising Multidisciplinary Areas for Research and Development

Conclusion. *High-field magnet science is intrinsically multidisciplinary.*

The construction of high-field magnets has always been motivated by the science that could be done with them, and in recent decades, physics, chemistry, biology, and medicine have all benefited from advances in magnet technology. For example, experiments done at the highest available magnetic fields have provided fundamental information about high-temperature superconductors, and instruments incorporating such magnets have enabled biochemists to study the structures and dynamics of large proteins. Physicians routinely use imaging instruments that contain high-field magnets to visualize the interiors of their patients. Since the performance of all such instruments generally improves with increasing field strength, the benefits that would accrue in these areas of science and medicine alone are more than enough to justify continued investment in magnet development. Even the technology that makes the construction of high-field magnets possible is cross-disciplinary. It is dominated by materials science and engineering, but several branches of physics contribute also.

Lying at the heart of much of the science done with magnetic fields today is the phenomenon of superconductivity. Investigation of superconductivity is a thriving area of materials science and condensed-matter physics, much of it now aimed at understanding how superconducting materials, especially high-critical-temperature superconductors, respond to magnetic fields. Scientific advances in this area will have widespread benefits because superconducting materials are used to build high-field magnets. The technology for making useful conductors out of high-

[3]For instance, both Bruker BioSpin in Germany and the high-field lab in Tsukuba, Japan, are developing 1-GHz NMR machines.

temperature superconductors as well as out of the recently discovered superconductor MgB_2 has advanced significantly in the last 2 years.

Major Construction Initiatives for the Coming Decade

Conclusion. *U.S. scientists will be unable to access a wealth of science opportunities if high magnetic field instrumentation is not provided at the Spallation Neutron Source and the nation's third-generation light sources. Also there are important issues relevant to the advancement of magnet technology that could be more efficiently addressed if the interested constituencies interacted more strongly, communicated with each other more fully, and coordinated their activities better.*

At present, there are no magnet construction initiatives under consideration in the high-field community that would require major new investments for their execution, aside from some application-specific magnet development plans such as those being considered in connection with the ITER fusion reactor, certain accelerators for high-energy physics, and high-field MRI. However, there are needs that will have to be addressed in the next decade if the United States is to remain internationally competitive in this important, fast-moving field.

RECOMMENDATIONS

The committee's recommendations follow below in order of priority, most important recommendations first.

In the course of its data gathering and deliberations, the committee made several observations relevant to NHMFL, which has facilities in Tallahassee, Florida; Gainesville, Florida; and Los Alamos, New Mexico. In making the following recommendation, the committee is not evaluating NHMFL. However, in responding to the charge "discuss and prioritize major new initiatives," the committee could hardly avoid identifying key components of the current laboratory system and highlighting areas that could be optimized. The committee bases the suggestions in its first recommendation on evidence presented in person at its open meetings, written testimony submitted in response to the call for input, impressions gathered during the site visit to NHMFL, and the direct experience of members who have used the facilities at NHMFL.

Recommendation 1a. *The United States should maintain a national laboratory that provides its scientific community access to magnets operating at the highest possible fields.*

A facility of this sort is essential to the vitality of many important scientific disciplines. NHMFL has successfully fulfilled this need for about a decade, and its activities have done much to foster the position that the United States currently enjoys in many areas of magnetic science and technology. The recent $10 million special appropriation by the state of Florida to address urgent infrastructure needs at the laboratory will help ensure that NHMFL can continue to perform this function.

> ***Recommendation 1b.*** *The quantity and the quality of the supporting instrumentation at NHMFL should be improved.*

At any high-field magnet laboratory, the capabilities of the devices available for controlling sample environments and measuring sample properties are almost as important as the field strengths of the magnets themselves. If more support personnel were available at NHMFL and more and better instrumentation were available, more effective work could be done and new research avenues would open up. The kinds of instrumentation that would be useful include setups to do magnetization, optical, electron-spin resonance, and pressure dependence measurements. The limited availability of the very low temperature (<10 mK) setup in Gainesville and the maximum field at which it can operate are other constraints.

> ***Recommendation 1c.*** *The operating schedule of NHMFL needs to be examined.*

For perfectly understandable budgetary and logistical reasons, NHMFL does not operate 24 hours a day, 7 days a week. But it is clear from information provided by NHMFL that its facilities are presently oversubscribed, a fact supported by the testimony of many of its users. NHMFL, in cooperation with NSF, should explore ways to maximize the return on capital invested in the national laboratory, such as longer hours of operation and flexible scheduling. The laboratory should undertake a cost-benefit analysis to identify the optimal balance between addressing user demand and the increased operating costs associated with longer hours of operation. The nation's synchrotron light sources and neutron-scattering centers, by contrast with NHFML, provide access 24 hours a day, 7 days a week, when in full operation, allowing visiting researchers to use their time at the facility to the best advantage. The trade-offs for expanding access to NHMFL need to be identified and weighed carefully, especially in constrained budget situations. The "extra" instrument time that would become available if NHMFL expanded its operations would enhance its scientific productivity and do much to alleviate the competition that now exists for access to its most unique magnets: the 33-T Bitter magnets and the 45-T hybrid magnet. Priority should also be given to improving the instrumentation available at its pulsed magnets so that some of the work now done on its DC magnets can be shifted to pulsed magnets.

> ***Recommendation 2.*** *New instruments for studying the neutron- and x-ray-scattering properties of materials in high magnetic fields should be developed in the United States.*

If history is any predictor of future success, instruments for making x-ray- and neutron-scattering measurements at high magnetic fields will deliver enormous scientific dividends. First-class science could be done with instruments that incorporate current state-of-the-art, high-field magnets; magnets that operate at even higher fields would be desirable but are by no means essential. The development of existing technology instruments is an obvious way to leverage the investments the nation has already made in high magnetic field science and technology and in x-ray- and neutron-scattering facilities.

The magnetic moment of neutrons makes them uniquely suited for investigating the magnetic properties of materials. The Spallation Neutron Source (SNS), which is under construction at Oak Ridge, will be the most powerful neutron source in the world when it begins operation, and it is the obvious place to start. The addition of high magnetic field instrumentation at SNS would provide U.S. scientists experimental capabilities unmatched elsewhere in the world. With these combined capabilities, researchers would be able to probe the dynamical properties of magnetic materials, the structure of magnetic moments in solids, and magnetic-field-induced phase transitions. It should also be noted that by improving the high magnetic field instrumentation at existing neutron facilities, capabilities could be obtained that are comparable to, or exceed, those now available in Europe. Because the high-field capability at the nation's light sources is also limited, the installation of high-field instrumentation at one of its third-generation light sources would be the logical second step in such a program. For economic reasons, the construction of state-of-the-art neutron or x-ray facilities at NHMFL seems an unattractive alternative.

It will be difficult to decide what kinds of high-field magnets should be provided at these scattering centers. On the one hand, the scientific opportunities that exist in the area of high-temperature superconductivity argue for installing magnets that generate the highest possible fields. On the other hand, mechanical, heating, cooling, and other considerations may require compromises. This committee was not asked to make more specific recommendations in this area, but it strongly recommends that these issues be pursued as soon as possible.

> ***Recommendation 3.*** *A consortium should be established to foster the development of magnet technology.*

In every scientific area the committee examined, magnets that produce fields higher than those available today would improve the reach of current research and, in many cases, bring new insights into unsolved problems. Order-of-magnitude increases in field strength cannot be expected, but 50 percent increases might be achievable, and they would be significant, as they have been all along. Improvements in field strength and, for pulsed magnets, field duration that are more than incremental will not be obtained unless many basic engineering and materials science problems are resolved, and it is these issues that the proposed consortium should address.

This recommendation is aimed primarily at the communities interested in high-field magnets and is motivated by the obvious advantages that might accrue to all if resources were pooled to solve common problems. The committee is open-minded about how this activity should be organized, but the objective is clear: to bring together scientists and engineers from *all* the communities working today on magnet technology, including the magnet engineers at NHMFL, academic researchers, the magnet designers in the high-energy physics and fusion communities, commercial vendors of superconducting magnets, including NMR and MRI systems, and manufacturers of advanced materials, such as high-strength materials and superconducting wire. The activities to be undertaken by the proposed consortium are sufficiently important to warrant federal support, but it would be wise for the membership to be international. Many of the communities that use high magnetic fields are already supranational in character, and several of the leading industrial magnet development companies are based overseas.

The sharing of information and resources within that larger community, which is now fragmented into components that communicate poorly, should accelerate the rate at which solutions are found to the problems they share. The committee proposes that the involved communities cooperate to establish a consortium whose objective would be to address the fundamental materials science and engineering problems that will have to be solved before the next generation of high-field magnets can be built. It is precisely because the needs of the many communities that use high magnetic fields overlap that the consortium approach seems appropriate. The committee envisions an effort that could range from something as small as a series of joint conferences or a virtual laboratory to a broad initiative formally facilitated by the Office of Science and Technology Policy that spans several agencies and has international involvement.

It is in the vein of recognizing common challenges that could be better addressed by coordinated efforts that the committee identifies several targets for magnet technology. As the underlying technology advances, some of the communities might decide to articulate more specific goals, such as the construction of magnets having the following specifications:

- *A 30-T superconducting, high-resolution magnet for NMR.* This technology would make it possible to build a 1.3-GHz NMR spectrometer for the study of large molecules. It would also permit the manufacture of magnets more suitable for condensed-matter physics and materials science laboratories than any available today.
- *A 60-T DC hybrid magnet.* Such a magnet would make it possible to do steady-state measurements at fields that can be accessed today only using pulsed field magnets.
- *A 100-T long-pulse magnet.* It will always be possible to reach higher fields with pulsed magnets than with DC magnets. A long pulse (i.e., ~100 ms) magnet of this field strength would enable a wide range of measurements at 100 T that cannot be made using magnets that operate for shorter times at the same field strength.

Other communities such as high-energy physics or fusion science might choose to focus on the materials science and engineering issues that surround high-volume production of the conductors needed for their magnet systems. The committee anticipates that it may be appropriate to build an engineering test facility (perhaps leveraging existing investments) to support the activity of such a consortium. It also acknowledges that this ambitious program could span more than a decade.

> ***Recommendation 4.*** *Agencies supporting high-field magnetic resonance research should directly support the development of technology and instrumentation for magnetic resonance and MRI.*

Without improvements in ancillary equipment, such as NMR probes, resonators, MRI coils, and radiofrequency electronics, the scientific benefits of higher magnetic fields will not be fully realized by the magnetic resonance community. Historically, the federal agencies supporting research that depends on magnetic resonance (NMR and electron paramagnetic resonance) and/or MRI have been reluctant to invest directly in this kind of technology. This attitude is short-sighted. Modest investments in NMR/EPR/MRI technology could result in improvements in instrument capability that have a large, beneficial impact on the quality and quantity of the data produced by many of the scientists these agencies support. (The committee recognizes that exploitation of the opportunities offered by the development of higher-field magnets will require concomitant advances in instrumentation and technique for nearly all applications in all disciplines, but in the area of magnetic resonance the need is particularly acute.)

Appendixes

A

Nobel Prizes for Research That Used or Significantly Affected the Development of High Magnetic Fields

Year	Field	Recipients	Citation
2003	Medicine	P. Lauterbur, P. Mansfield	Magnetic resonance imaging
2003	Physics	A. Abrikosov, V. Ginzburg, A. Leggett	Type II superconductors, superfluidity
2002	Chemistry	K. Wuthrich	NMR spectroscopy of biological macromolecules in solution
1998	Physics	R. Laughlin, H. Stormer, D. Tsui	Fractional quantum Hall effect, theory and experiment
1996	Physics	D. Lee, D. Osheroff, R. Richardson	Superfluidity of helium-3
1991	Chemistry	R. Ernst	High-resolution Fourier transform and 2D NMR
1987	Physics	J. Bednorz, K. Muller	High-T_c superconductivity
1985	Physics	K. von Klitzing	Quantum Hall effect
1982	Physics	K. Wilson	Critical phenomena, phase transitions
1977	Physics	P. Anderson, N. Mott, J. van Vleck	Theory of magnetic and disordered systems
1973	Physics	L. Esaki, I. Giaever, B. Josephson	Superconducting tunnel junctions
1972	Physics	J. Bardeen, L. Cooper, J. Schrieffer	Theory of superconductivity
1970	Physics	L. Neel	Antiferromagnetism, ferrimagnetism
1955	Physics	P. Kusch	Precision measurement of the electron's magnetic moment
1952	Physics	F. Bloch, E. Purcell	NMR in condensed matter
1944	Physics	I. Rabi	NMR of isolated atoms and molecules, using molecular beams
1943	Physics	O. Stern	Magnetic moment of the proton
1922	Chemistry	F. Aston	Mass spectrograph
1913	Physics	K. Onnes	Superconductivity

B

High-Field Magnet Facilities Around the World

This appendix summarizes the high magnetic field facilities and activities available around the world. The information was distilled from a combination of personal communication with facility staff and a review of the literature.[1] It is not meant to be exhaustive but is meant to give the reader a sense of the scope of activity in the field at home and abroad.

Table B.1 lists the details of the equipment available at facilities dedicated to high magnetic field science. Table B.2 briefly describes the high magnetic field services available at other large research facilities such as synchrotron and neutron sources.

High magnetic field labs fall into roughly three classes based on their capabilities (the breakdown below takes into account near-term planned capabilities):

- Local facilities (affordable by many research groups)
  —15-25 T DC superconducting magnets
  —60-80 T pulsed-field magnets of 5-20 ms duration
  —100-200 T destructive pulsed-field installations of 5 μs duration

[1]The committee gratefully acknowledges the work of F. Herlach and N. Miura as published in the introduction to *High Magnetic Fields: Science and Technology, Vol. 1*, World Scientific Publishing, Co.: Singapore, 2003, which they also edited. Their survey of magnet laboratories served as an important cross-reference for the committee's research.

TABLE B.1 Dedicated High Magnetic Field Facilities

Location	Name	DC Fields and Magnets	Pulsed Fields and Magnets
Australia			
Sydney	National Magnet Laboratory		60 T – 25 ms – 400 kJ – 22 mm
Austria			
Vienna	Austromag		40 T – 10 ms – 20 kJ – 25 mm
Belgium			
Leuven	Pulsed Field Facility		55 T – 20 ms – 300 kJ – 18 mm 70 T – 8 ms – 260 kJ – 10 mm
China			
Hefei	High Magnetic Field Facility	8 + 3 magnets: 15 T – 50 mm (resistive) 20.2 T – 32 mm (hybrid) [26 T (hybrid)] 8 T – 54 mm (sc)	[45 T]
England			
Bristol	Wills Physics Laboratory	20 T	60 T – 50 ms – 12 mm
Oxford	Nicholas Kurti Magnetic Field Laboratory	21.5 T – 40 mm (sc)	60 T – 10 ms – 150 kJ – 12 mm 50 T – 20 ms – 250 kJ – 20 mm
France			
Grenoble	Grenoble High Magnetic Field Laboratory	7 + 15 magnets: 30 T – 50 mm (resistive) 40 T – 34 mm (hybrid) 18 T – 50 mm (sc)	
Toulouse	National Laboratory of Pulsed Magnetic Fields		73 T – 10 ms – 8.5 MJ – 15 mm 61 T – 150 ms – 1.25 MJ – 11 mm 58 T – 282 ms – 3.30 MJ – 19 mm 45 T – 500 ms – 8.1 MJ – 120 mm
Germany			
Berlin	Humboldt High Magnetic Field Center		51 T - 3.5 ms – 42 kJ – 10 mm 60 T - 8.1 ms – 400 kJ – 18 mm
Braunschweig	High Field Magnet Laboratory	4 + 1 magnets: 18 T – 32 mm (resistive) 11 T – 37 mm (sc)	27 T – 40 ms – 20 kJ – 12 mm
Dresden	High Field Laboratory		50 T – 13 ms – 450 kJ – 24 mm 40 T – 120 ms – 1,100 kJ – 24 mm [100 T – 10 ms – MJ – 20 mm] [70 T – 100 ms – MJ – 24 mm] [60 T – 1,000 ms – MJ – 50 mm]
Frankfurt	Institute of Physics		36 T – 600 ms – 800 kJ – 22 mm 50 T – 24 ms – 390 kJ – 20 mm
Ireland			
Dublin	Trinity College		26 T – 700 ms – 200 kJ – 26 mm

continued

TABLE B.1 Continued

Location	Name	DC Fields and Magnets	Pulsed Fields and Magnets
Italy			
Parma	Institute of Electronic and Magnetic Materials		60 T – 60 ms – 760 kJ – 27 mm 50 T – 65 ms – 880 kJ – 15 mm
Japan			
Kashiwa	MegaGauss Laboratory	20 T (sc)	50 T – 50 ms – 22 mm 50 T – 20 ms – 200 kJ – 20 mm 50 T – 18 ms – 200 kJ – 20 mm
Kobe	High Field Magnet Laboratory	10 T (sc)	30 T – 11 ms – 24 kJ -15.4 mm 36 T – 9 ms – 34 kJ - 16.5 mm
Okayama	Okayama University		35 T – 3 ms – 60 kJ – 17 mm 30 T – 1 ms – 60 kJ – 31 mm
Osaka	Kyokugen		60 T – 7 ms - 240 kJ – 18 mm 70 T – 7 ms – 240 kJ – 10 mm 60 T – 22 ms – 570 kJ – 18 mm
Sendai	High Field Laboratory for Superconducting Materials	3 + 12 magnets: 20 T – 32 mm (resistive) 31 T – 32 mm (hybrid) 20 T – 52 mm (sc)	40 T – 10 ms – 70 kJ – 17 mm 30 T – 5 ms – 45 kJ – 22 mm
Tsukuba	Tsukuba Magnet Laboratory	4 + 9 magnets: 29 T – 32 mm (resistive) 35 T – 32 mm (hybrid) 23 T – 13 mm (sc)	48 T – 20 ms – 500 kJ – 16 mm 50 T – 15 ms – 300 kJ – 20 mm
Netherlands			
Amsterdam	University of Amsterdam	40 T	40 T – 1,500 ms – 6 MW – 20 mm
Nijmegen	High Field Magnet Laboratory	6 + 3 magnets: 33 T – 32 mm (resistive) [40 T – 32 mm (hybrid)] 33 T – 40 mm (sc)	60 T – 5 ms – 2 MJ – 23 mm
Poland			
Wroclaw	International Laboratory of High Magnetic Fields and Low Temperatures	3 + 3 magnets: 20 T – 25 mm (resistive) 15 T – 20 mm (sc)	47 T – 35 ms – 70 kJ – 10 mm
Portugal			
Porto	Centro de Fisica at the University of Porto		25 T – 2,000 ms – 600 kJ – 30 mm
Russia			
Moscow	Kurchatov Institute	18.3 T – 28 mm (resistive) 24.6 T – 28 mm (hybrid) 17.7 T – 40 mm (sc)	55 T – 9.5 ms – 180 kJ – 15 mm
Sarov	VNIIEF		2,800 T – 5 mm
St. Petersburg	Ioffe Institute and State Technical University		35 T – 22 mm

TABLE B.1 Continued

Location	Name	DC Fields and Magnets	Pulsed Fields and Magnets
Spain			
Zaragoza	Long Pulse Magnet Facility		31 T – 2,000 ms – 1,200 kJ – 30 mm
United States			
Cambridge, Mass.	Francis Bitter Magnet Laboratory	24 T – 32 mm (resistive) 35.2 T – 32 mm (hybrid)	65 T – 7.3ms – 110 kJ – 13 mm 60 T – 9.4ms – 180 kJ – 20 mm
Los Alamos, N. Mex.	National High Magnetic Field Laboratory	1 + 3 magnets: 20 T – 52 mm (sc)	60 T – 35 ms – 565 kJ – 15 mm 50 T – 350 ms – 1,400 kJ – 15 mm
Tallahassee and Gainesville, Fla.	National High Magnetic Field Laboratory	10 + 33 magnets: 33 T – 32 mm (resistive) 45 T – 32 mm (hybrid) 20 T – 52 mm (sc)	
Worcester, Mass.	Clark University		51 T – 80 ms – 296 kJ – 18 mm 42 T – 150 ms – 180 kJ – 18 mm

NOTE: The column listing DC magnets shows the total number of continuous-field magnets available, separated into resistive/hybrid plus (+) superconducting magnets and then lists parameters for a representative of each type of magnet available (resistive, hybrid, and superconducting). Magnets currently under construction have parameters enclosed by brackets ([ ]).

- Regional or national facilities (several worldwide)
  —25-35 T DC resistive magnets (10-20 MW)
  —40 T DC hybrid magnets (20 MW)
  —40-60 T pulsed-field magnets of 10-1,000 ms duration
- International facilities (a few worldwide)
  —35-40 T DC resistive magnets (20-40 MW)
  —50 T DC hybrid magnets (40 MW)
  —80-100 T pulsed-field magnets of 0.1-1.0 s duration

AUSTRALIA

The Australian National Magnet Laboratory (NML), established in 1991 at the University of New South Wales in Sydney, provides Australian scientists interested in condensed-matter physics facilities for doing cutting-edge research at high magnetic fields and low temperatures. Experiments involving electrical, optical, and far-infrared studies of low-dimensional semiconductor nanostructures, organic layered structures, and conventional and high-T_c superconductors are all possible at this facility. NML offers pulsed magnetic fields up to 60 T at temperatures down to 70 mK for research in correlated-electron phenomena.

TABLE B.2 Description of Major User Facilities Not Specifically Dedicated to the Pursuit of High Fields, Inside and Outside the United States

	Scope of Facility			Use and Availability of High Magnetic Fields
Facility	Users per Year	Operating End Stations	Experiments per Year	Fraction of Experiments Using >5T
United States				
Synchrotron sources				
National Synchrotron Light Source, Brookhaven National Laboratory	2,206	87 (94 max)	1,145	0.50%
Advanced Photon Source, Argonne National Laboratory	2,700	41 (68 max)	2,172	"few %"
Advanced Light Source, Lawrence Berkeley National Laboratory	1,662	36	200	1 series out of 200
Stanford Synchrotron Radiation Laboratory, Stanford University	1,000	27	1,000	18 out of 900
Cornell High Energy Synchrotron Source, Cornell University	500	12	>100	Rare

Use and Availability of High Magnetic Fields		Future Plans for High Magnetic Fields		
Existing Capabilities	How Owned and Maintained	Foresee Greater Need	Field Strength for New Opportunities	Availability of Support
15 T, vertical, 25 mm D sample 13 T, vertical, 20 mm D sample 10 T, horiz., split pair, 25 mm 9 T, parallel thru bore, 3 in D 5 T split pair, vertical, 3.5 cc	By PIs; available to users by outside proposal or collaboration. Limited support.	Yes. Demand is there, but instruments and support are not.	25 T, large bore, 3-300 K sample temperatures. "No field is high enough."	Requires additional resources. Present arrangement will not allow widespread use.
4 T, horiz., 1 cubic inch 7 T, horiz. 10 T vertical 4 T vertical 7 T horiz. 7-8 T horiz.	Owned and maintained by individual teams. Accessible to general users by proposal.	Workshop upcoming. Users believe that 10-20 T needed steady. Pulsed not well matched.	15-17 T	Not sure of support level required, but it is not available now.
6 T	Magnet is owned and maintained by PI.	Yes, but no proposals have yet been declined.	Not specified. Higher is better, but size and complexity are traded off.	No. Would require additional resources.
13 T, split coil, 2 cm D	Owned and operated by local user group. General users by proposal.	New field at SSRL but growing. Need likely to grow.	20 T (study of correlated electron materials).	No. Additional resources required.
None	NA	Users have not requested. Will respond to demand	No plans.	No. Neither space nor staff.

continued

TABLE B.2 Continued

	Scope of Facility			Use and Availability of High Magnetic Fields
Facility	Users per Year	Operating End Stations	Experiments per Year	Fraction of Experiments Using >5T
Neutron sources				
High Flux Isotope Reactor, Oak Ridge National Laboratory	51 now; 750 at end of upgrades	3 now; 15 planned	58 now; 450 predicted	10%
Intense Pulsed Neutron Source, Argonne National Laboratory	229	15	370	9%, <5 T
Spallation Neutron Source (now under construction), Oak Ridge National Laboratory	1,000 predicted in 2008	(10 in 2008)	400 predicted for 2008	20% anticipated
Lujan Center at Los Alamos Neutron Science Center, Los Alamos National Laboratory	280	8 user; 3 limited access	221	3%
NIST Center for Neutron Science, National Institute of Standards and Technology	800	24 (18 scattering)	1,100	5%

Use and Availability of High Magnetic Fields	Future Plans for High Magnetic Fields			
Existing Capabilities	How Owned and Maintained	Foresee Greater Need	Field Strength for New Opportunities	Availability of Support
7 T vertical, 30 mm; 1.5-300 K 5 T vertical, 50 mm; 1.5-300 K 5 T horiz., 40 mm; 1.5-300 K	Facility technicians maintain. Scientific staff help fill, mount samples.	Substantial demand from users (cf Europe).	15 T to start; go to 20 T when possible.	Would require additional resources.
None at this time	Facility technical staff and scientific staff. Available by proposal.	Yes; 7 T on order	Any field > current opens up new opportunities.	Would have to be operated by facility scientific personnel.
Plan: 2-9 T vertical; 50 mm D, 20 mm high 1-7 T horiz., 100 mmD, 20 mm high 2-5 T large bore (100 mm) 1-14 T vertical; 1 "XYZ"	Facility plans to have 9-person staff with help from scientists.	Yes	30 T would open up new opportunities.	No. Very high field would require more staff.
11 T, vertical, split coil, 35 mm D sample, 2-300 K	Beam line scientists. Accessed by proposal.	Yes	Planning (with NHFML) 30-T pulsed, 50-mm bore	No. Require more staff if successful.
7 T, vertical; 71 mm D; 4.1-300 K 9 T vertical; 28 mm D; 0.3-300K 9 T horiz., 28 mm D; 2.7-300 K 11.5 T vertical; 50 mm D; 0.02-4 K (13 T vertical; 25 mm D; 0.3-300 K)[a]	Facility owned and operated; technicians and beamline scientists. Access by proposal with NIST staff.	Yes, current limit is staff to maintain	>15 T	No. Cannot support current inventory.

continued

TABLE B.2 Continued

	Scope of Facility			Use and Availability of High Magnetic Fields
Facility	Users per Year	Operating End Stations	Experiments per Year	Fraction of Experiments Using >5T
Europe and Japan[b]				
Synchrotron sources				
European Synchrotron Research Facility, Grenoble, France	2,206	50 (64 possible)	1,116	5-6%
Neutron sources				
Berlin Neutron Scattering Center (BENSC), Hahn-Meitner Institute, Berlin, Germany	600	20	300	>25%
Institut Laue Langevin, Grenoble, France	>2,000	40	750	
ISIS, UK Rutherford Appleton Laboratory	1,700	19 neutron 4 muon	800	8%
FRM-II, Munich, Germany	Just beginning operation.	20	No data	No data, but expect 6-7 instruments.

Use and Availability of High Magnetic Fields	Future Plans for High Magnetic Fields			
Existing Capabilities	How Owned and Maintained	Foresee Greater Need	Field Strength for New Opportunities	Availability of Support
7 T, 1.5 K 7 T, low temperature	Facility owned and operated.	Yes, are planning for 10 T magnet in 2004. Future 20 T pulsed magnet		
Two 14.5 T, 1.5-300 K, 20 mm D, vertical 7 T, 1.5-300 K, 50 mm D, vertical 5 T, 1.5-300 K, 50 mm D, vertical 5.5 T, 1.5-300 K, 50 mm D, vertical 6 T, 1.5-300 K, 50 mm D, horiz. 4 T, 1.5-100 K, 40 mm D, horiz. 2 dilution inserts, 25 mK	Facility owned and operated. User access by proposal.	Yes. Have proposal for 40 T magnet to study HTS and quantum critical phenomena.	20-40 T	
All vertical field, 1-300 K 5.5 T, 300 cc 6 T, 260 cc 6 T, 360 cc 7 T, 50 cc 10 T, 7 cc 10 T, 150 cc 12 T, 80 cc 15 T, 40 cc Some capable of dilution insert	Some facility staff, some research groups. Mostly operated by facility. Accessed by proposal.	Yes. Resources limit ability to pursue.		Limited for new capability.
7.5 T, vertical, 50 mm D 10 T, vertical, 50 mm D	Facility owned and operated. Access through user proposal.	Yes, if resources available.	At least 20+ T to match local user capabilities.	More capability will require more resources to support.
15 T vertical, 50 mm D, low temperature 10 T vertical, 100 mm D, warm	Facility owned and operated. Access by proposal or collaboration.	Yes	25 T or more.	Require staff beyond 20 T.

continued

TABLE B.2 Continued

	Scope of Facility			Use and Availability of High Magnetic Fields
Facility	Users per Year	Operating End Stations	Experiments per Year	Fraction of Experiments Using >5T
JRR-3M, JAERI, Tokai, Japan	350	11	113	10%
Research Reactor Institute, Kyoto, Japan	30	7	30	15%

[a]Being repaired if possible.
[b]Selected facilities.

AUSTRIA

The Austromag facility at the Institute for Experimental Physics at the Vienna University of Technology is a small laboratory that was established in 1975 to develop pulsed magnetic field instrumentation. The instruments available today can generate 40 T fields in long pulses (1 s), and work is under way to modify them so that fields up to 70 T can be produced. The design calls for superimposing a short (about 5 ms) 30-T pulse on longer 40-T pulses. The device will be used primarily to study the magnetic properties of materials over broad temperature ranges.

BELGIUM

The Pulsed Field Facility at the Laboratory of Solid State Physics and Magnetism at the Catholic University in Leuven, Belgium, operates a pulsed-field facility that has five parallel measuring stations. User magnets are designed to work reliably up to 70 T with a pulse duration about 20 ms. The Leuven group has been active in coordinating the European Community-sponsored initiative for the development of strong conductor materials and associated coil designs. A new type of pulsed magnet with optimized fiber composite reinforcement was developed at Leuven, resulting in record fields of 75 T. Research is being done on

Use and Availability of High Magnetic Fields	Future Plans for High Magnetic Fields			
Existing Capabilities	How Owned and Maintained	Foresee Greater Need	Field Strength for New Opportunities	Availability of Support
10 T vertical, 150 mK-300 K, 30 mm D 5 T vertical, 20 mK-6 K, 76 mm D 5 T horizontal, 2-300 K 13.5 T vertical, 150 mK-300 K, 30 mm D	By facility. Access by collaboration after proposal.	Yes	15-20 T 50 T breakthrough. Pulsed at 20T?	Require additional for new opportunities.
5 T vertical	By beam line. Access by collaboration.	Future uncertain. Shutdown likely.		

high-temperature superconductors, quantum dots, the normal state properties of high-temperature superconductors, the magnetic properties of spin-Peierls compounds, magnesium dioxide CMR materials and iron oxide garnet compounds, and the photoluminescent properties of low-dimensional semiconductor structures such as self-assembled quantum wells.

CHINA

The Chinese Academy of Sciences' Institute of Plasma Physics maintains a high magnetic field facility in Hefei focused on developing technologies for its tokamak fusion program, the Experimental Advanced Superconducting Tokamak. Research into the physical properties, processes, and mechanisms of condensed matter at high magnetic fields is being conducted at the institute in cooperation with scientists from the Institute of Solid State Physics and the University of Science and Technology of China. Additional research is being conducted on the effects of high magnetic field on chemical reactions and biological processes and on industrial applications of high-field magnet technology. Using magnets and power supplies from Grenoble, the facility provides 11-T fields to users and it plans to upgrade in stages to hybrid magnets of 23 T and 26 T. A long-pulse magnet of 45 T is also under development.

ENGLAND

Oxford University

The Nicholas Kurti Magnetic Field Laboratory at Oxford University, which opened in April 2002, builds on the expertise of the Clarendon Magnet Laboratory, which dates back to the 1930s and has a long history of research with water-cooled and hybrid magnets. It currently maintains several DC superconducting magnets (18 T and 21.5 T), which will be upgraded using high-T_c inserts. Clarendon and the group at Leuven formed the nucleus of the European collaboration EUROMAGTECH, which has been working on pulsed-field magnet design and materials development. The new Kurti Laboratory houses a three-station, pulsed-field laboratory, which will generate nondestructively the highest fields available in the U.K. (>60 T). The devices at the Kurti Laboratory will be used by several research groups. Research will be done on transport current superconductivity and in areas such as biophysics, nanostructures, and engineering technology. Magnetization of bulk high-temperature superconductors is another focus.

University of Bristol

In 1988, the Wills Physics Laboratory at the University of Bristol acquired some pulsed magnets, superconducting coils, and dilution refrigerators. The facility there now hosts two 52-T pulsed magnet devices and two continuous magnets with strength over 20 T. Research focuses on heavy fermion compounds, superconductors, and mesoscopic systems.

FRANCE

Grenoble

The Grenoble High Magnetic Field Laboratory (GHMFL) is a French-German laboratory jointly funded by the Centre National de la Recherche Scientifique and the Max-Planck Gesellschaft (MPG) and run partly as a user facility. It operates seven different magnets that produce DC magnetic fields up to 30 T, and a new 40-T hybrid magnet is under construction. A 60-T pulsed coil is also available. Along with the European Synchrotron Radiation Facility, GHMFL is developing instrumentation that will enable scientists to use x rays to study materials in high pulsed magnetic fields. Since 1992, GHMFL has been open to users from the rest of the European Union and associated states on the same basis as French and MPG users. Research is also performed by a resident scientific staff. Primary research activities include work with 2D electron systems, nanostructures, bulk semiconductors, and there is some work with biological and inorganic systems using electron paramagnetic resonance.

Toulouse

The Laboratoire National des Champs Magnétiques Pulsés (LNCMP) in Toulouse is the European leader in long-pulse magnetic fields, offering field strengths up to 62 T and pulse durations up to 0.2 s. The three missions of the laboratory are (1) to serve as a user facility for the French and European scientific communities, (2) to develop techniques for pulsed magnetic field production, and (3) to develop innovative research activities in pulsed magnetic fields. The primary research interests using the high fields are high-T_c superconductors, organic conductors, quantum magnetism, and low-dimensional (semi)conductors. LNCMP has seven pulsed magnets ranging in field strength from 38 to 77 T and one 16-T superconducting magnet; its high-voltage generator can store 14 MJ at 24 kV. The lab is planning to build three more pulsed magnets (the lab has 10 magnet sites) and to upgrade its 38-T magnets to higher fields or bigger magnet bores—for example, to make a 14-MJ magnet with 700 ms duration. The European ARMS project magnet, which will use a two-coil system to deliver 80 T, will also be sited at Toulouse.

GERMANY

Berlin

Humboldt University in Berlin houses the Megagauss Facility, which specializes in the production of high pulsed fields of up to 300 T. It was the first laboratory to employ the single-turn coil technique systematically for megagauss experiments. The lab focuses on the generation and application of strong magnetic fields using semidestructive single-turn-coils, although nondestructive pulsed magnets (50 T, 60 T) and continuous-field magnets (18 T, 20 T) are also available. The primary area of research is semiconductor spectroscopy in high magnetic fields to characterize microscopic parameters that cause conductivity, especially DC-magnetotransport and FIR-magnetospectroscopy in low-dimensional materials.

Braunschweig

Established in 1963, the High-Field Magnet Laboratory at the Braunschweig Technical University provides research opportunities for faculty, students, and visiting scientists. The facility includes four Bitter magnets (up to 18.T with a

include magnetospectroscopy and high-field NMR (>10^{-6} resolution at 30 T, 1.3 GHz).

POLAND

The International Laboratory of High Magnetic Fields and Low Temperatures (ILHMFLT), in Wroclaw, was founded n 1968. The participating organizations are the Bulgarian Academy of Sciences, the Polish Academy of Sciences, and the Russian Academy of Sciences, and some support is provided by the European Union. ILHMFLT includes three continuous-field magnets operating at 10 T, 15 T, and 20 T, three DC magnets that are being tested, and 40-T, 50-T, and 60-T magnets that deliver pulses of 0.1-1.0 s. Research is being carried out on techniques for generating high magnetic fields, on superconductors, on magnetics, and on the electronic structure of metals, their compounds, and conductive organic substances. A specialty of the lab is research under high pressure.

PORTUGAL

The Centro de Fisica at the University of Porto maintains a pulsed-magnet installation that was designed and constructed at Toulouse and provides up to 26 T over 2 s at 4 to 300 K. The primary research areas are magnetic thin films and nanostructured materials.

RUSSIA

Moscow

The Division of Superconducting Magnets and Cryogenics at the Institute of Superconductivity and Solid State Physics at the Russian Research Center (a.k.a. the Kurchatov Institute) has a long tradition of work with pulsed magnets (up to 55 T) as well as with superconducting magnets. Over the last several years, the division has developed, manufactured, and tested superconducting magnets for use at other institutions, including devices for nonneutral plasma studies at Berkeley, the Chinese synchrotron source, and the tokamak fusion program in India. Research covers various applications of large-scale superconductivity such as superconducting wires, cables, conductors, and other specific features of HTS magnet technology.

Sarov

The Institute for Fundamental and Applied Physics Research at the All-Russia Research Institute of Experimental Physics (VNIIEF) in Sarov specializes in explosive flux compression techniques to achieve pulsed high magnetic fields. Efforts at the Institute since 1952 have addressed the conversion of explosion energy into magnetic energy and the use of this concept to design magnetic-cumulation or explosive-magnetic generators. The magnetic-cumulation generator is a device that can produce magnetic fields as high as 1,000 T in an active volume 9 mm in diameter and 100 mm long. The world record of 2,800 T was achieved with a 170-kg explosive-driven, magnetic flux compression device. Research with these magnets ranges from weapons applications to detailed magnetic analyses of molecules.

St. Petersburg

Established in 1964, the laboratory of the Ioffe Institute and State Technical University specializes in magnetotransport measurements for samples with high resistance using pulsed fields of up to 35 T. At the nearby State Technical University, research on high-field production using destructive means is pursued. Capacitor-driven single-turn coils have been developed extensively; fields of over 70 T have been achieved.

SPAIN

Zaragoza

The long-pulse magnet facility of the University of Zaragoza, part of the Consejo Superior de Investigaciones Cientificas, was established in 1994 and produces fields up to 31 T, with a pulse duration of 2.2 s. The inner bore of the helium-4 cryostat is 22.5 mm. Research topics include colossal magnetoresistance, magnetoelastic stresses and strains, and magnetic thin films.

UNITED STATES

Cambridge, Massachusetts

The Francis Bitter Magnet Laboratory at the Massachusetts Institute of Technology is one of the oldest pulsed-field laboratories in the world. This laboratory opened up many of the areas of magnet science and technology that are still active today, ranging from the fractional quantum Hall effect to far-infrared cyclotron resonance and the construction of high-performance magnets using high-strength conductors. Its long-pulse magnets, which operate at 60 T and 65 T, are still in use. The National Magnet Laboratory was located at the Francis Bitter lab from 1960 to 1995. Since the opening of NHMFL, the focus of the Cambridge facility has shifted to magnetic resonance. It now offers users 25 NMR and EPR spectrometers, and two new 900-MHz spectrometers are under development.

Los Alamos, New Mexico

The Pulsed Field Facility at Los Alamos National Laboratory is one of the three campuses of NHMFL, the other two being at Florida State University in Tallahassee (continuous fields, magnetic resonance, and general headquarters) and the University of Florida in Gainesville (ultralow temperatures at high magnetic fields). NHMFL is sponsored primarily by the National Science Foundation, Division of Materials Research, with additional support from the state of Florida and the U.S. Department of Energy. The Pulsed Field Facility is the only facility of its kind in the country because of its 1.4-GW motor generator. The facility operates both short- and long-pulse magnets and a broad library of programmable waveform shapes in between; a record 80 T was achieved in 1999. A unique 100-T nondestructive pulsed hybrid magnet is under construction. The flagship 60-T long-pulse magnet failed catastrophically in 2000 but is being returned to operation. Users also have limited access to the high fields generated using destructive pulsed magnets. Supporting instrumentation at the LANL facility is also extensive and state of the art.

Richland, Washington

High-field NMR instruments and high-performance ion cyclotron resonance (ICR) instruments are available at a national user facility housed in the Pacific Northwest National Laboratory's William R. Wiley Environmental Molecular Sciences Laboratory (EMSL). EMSL was organized in 1997. The NMR facility includes 900-, 800-, and 750-MHz devices (along with nine other NMR spectrometers) and an EPR spectrometer. The science topics investigated by users include structural and functional genomics and biological imaging as well as instrumentation development. The mass spectrometry facility incorporates several ion trap mass spectrometers and four Fourier transform ion cyclotron resonance mass spectrometers, including an 11.5-T instrument and a new 9.4-T instrument. Much of the work conducted using these spectrometry facilities concerns proteomics, cell signaling, high-molecular-weight systems, and cellular molecular machines.

Tallahassee and Gainesville, Florida

The NHMFL was established in 1990 with support from the National Science Foundation and the state of Florida and is a collaboration between the University of Florida, Florida State University, and Los Alamos National Laboratory. The laboratories in Florida provide users access to continuous field magnets, including a 45-T hybrid magnet; they constitute the world's largest magnet laboratory. In addition to an active research program in static high magnetic fields, there is an extensive program in NMR, ICR, EMR, and other advanced high-field magnetic resonance research. The continuous-field magnet facility, the Center for Interdisciplinary Magnetic Resonance, and the geochemistry facilities are located on the main campus in Tallahassee; the high B/T (magnetic field/temperature) and the magnetic resonance/imaging spectroscopy facilities are located in Gainesville.

Worcester, Massachusetts

The laboratory at Clark University focuses on organic conductors and the effects of high magnetic fields on their properties. On-campus magnets include 5-T continuous-field magnets and a 50-T (11-ms) pulsed-field facility. The focus of the instrumentation available is cryogenic equipment that can be used in high pulsed field environments, such as plastic cryostats.

C

Glossary

antiferromagnetism: In substances known as antiferromagnets, the magnetic moments of adjacent atoms tend to line up antiparallel, yielding an ordered state with no net magnetic moment. Consequently, such materials display almost no response to an external magnetic field at low temperatures.

band structure, bandgap: Theory describing the collective organization and interaction of atoms (and notably their valence electrons) in a solid. The band structure of a solid is the continuous range of states with different energies that are filled by the charge carriers in an extended solid. In insulators and semiconductors, there is a bandgap that separates the last filled state from the first excited state, unoccupied at zero temperature. The electrical properties of a material at room temperature can be influenced by the (temperature-dependent) population of charge carriers near the bandgap. For instance, materials without a bandgap (i.e., metals) or with a bandgap comparable to the thermal energy (i.e., semiconductors) are typically conductors, while materials with a very wide bandgap are typically insulators.

Bitter magnet: Design for DC resistive magnets invented by Francis Bitter in the late 1930s. Bitter's design meets the high conductivity and cooling requirements of high-field resistive magnets using perforated copper plates that are sandwiched between insulating layers, a small region of electrical contact being allowed between plates so that current can flow from one plate to the next. Electrical current flows through the resulting copper spiral, and coolant flows through perforations in the conductors, which are aligned vertically.

bore, magnet: Inner diameter of a cylindrical magnet where the magnetic field is generated. The bore of a magnet constrains the volume available for experimental use.

coil, magnet(ic): Electric current in most electromagnets passes through coils of wire. Since the coils of all such magnets are their active component, the terms coil and magnet are often used as synonyms.

coherence length: Characteristic scale of a Cooper pair in a superconducting material. The coherence length effectively represents the longest distance over which the two electrons of the Cooper pair act in tandem and is typically on the order of 1.5 nm for high-field materials.

Cooper pair: Entity believed to explain the superconductivity of many materials. A Cooper pair consists of two electrons that are paired together into a new state with zero net charge and angular momentum. Below the superconducting transition temperature, Cooper pairs form a condensate—a macroscopically occupied single quantum state—in which current flows without resistance.

conductor: Material such as Cu or Al in which charge carriers can move under the influence of an electrical voltage. Unlike superconductors, conductors have finite, nonzero resistance.

correlated electron systems: Also strongly correlated electron systems; a many-particle system in which strong interactions between electrons play a crucial role in determining fundamental properties. Electronic correlations can cause striking many-body effects like superconductivity, electronic localization, magnetism, and charge ordering, which cannot be described using the simpler independent particle picture. These properties and dynamics arise from the collective interactions of the electrons with one another.

critical current density (J_c): At a certain temperature, the maximum electrical current density that a superconductor can carry before it quenches and enters the normal state. In general, as the current flowing through a superconductor increases, the T_c (see below) will usually decrease.

critical field (H_c): At zero applied current, the maximum magnetic field (at a given temperature) that a superconductor can transport before it quenches and returns to a nonsuperconducting state. Typically, a higher T_c (see below) is associated with a higher H_c.

cryogenically cooled probe: Device installed in the bore of an NMR magnet that carries the samples to be studied as well as the electronics necessary both for perturbing the orientation of nuclear spins in samples and for detecting the consequences of those perturbations electromagnetically. Probes may include additional devices for controlling the sample environment. In a cryogenically cooled probe, in order to improve signal to noise, electronic components are cooled to liquid helium temperatures, which minimizes shot noise.

cyclotron: Device for experimental particle physics that uses an oscillating electric field to accelerate charged particles and a magnetic field to control particle trajectories. Charged carriers of all kinds follow cyclotron-like spiral trajectories in magnetic fields. An important quantity in high magnetic field studies is the cyclotron frequency, which is a measure of the density of mobile carriers in doped semiconductors and metals.

DC magnet: Steady-state magnet. DC stands for direct current, meaning that the flow of current in the magnet's coils is constant in time.

electromagnet: Device designed to generate a magnetic field by having electric current passed through it.

electron paramagnetic resonance (EPR): Electron paramagnetic resonance (EPR), also known as electron spin resonance (ESR) or electron magnetic resonance (EMR), is the resonant absorption of microwave radiation by paramagnetic ions or molecules with at least one unpaired electron spin in the presence of a static magnetic field. It has a wide range of applications in chemistry, physics, biology, and medicine. For example, it may be used to probe the static structure of solid and liquid systems and is also very useful in investigating dynamic processes.

ferromagnetism: In substances known as ferromagnets, the magnetic moments of adjacent atoms tend to line up in parallel, yielding an ordered state that has a macroscopic magnetic moment. Once the magnetic domains in a ferromagnetic substance have become aligned by a small field, the magnetic moment of the bulk material will persist even in the absence of an external magnetic field, a property unique to ferromagnets. Consequently, ferromagnetic materials can be used to make permanent magnets that deliver fields as large as 1-2 T. Elements such as iron, nickel, and cobalt are ferromagnetic at room temperature.

fusion: Nuclear reaction in which nuclei combine to form more massive nuclei with the simultaneous release of energy.

gauss (G): Unit of measure for magnetic field strength in the cgs system of units. Earth's magnetic field is about 0.5 G. One G is equal to 0.0001 tesla (T), the mks unit of magnetic field.

hybrid magnet: In a hybrid magnet system, resistive and superconducting magnet technologies are combined. The superconducting magnet takes the place of the outer portion of the resistive coil. The resistive portion operates as an insert to the superconducting magnet and produces the portion of the field that exceeds the critical current and field limits of the superconducting magnet.

hybrid magnet, series-connected: Hybrid magnet system where the current supplied to the resistive insert travels first through the superconducting outer magnet.

ion cyclotron resonance (ICR) mass spectroscopy (ICRMS): Method for precisely measuring the mass of a collection of ions originating from the chemical dissociation of complex molecules and solids; it depends on cyclotron resonance. Ions with a range of mass-to-charge ratios are exposed to a high-frequency electric field in the presence of a constant magnetic field perpendicular to the varying electric field. Maximum energy is gained by the ions that satisfy the cyclotron resonance condition and that can be separated on that basis from ions that have only a slightly different mass-to-charge ratio.

J_c: See *critical current*

linewidth: Energy resolution of a feature in an experimental measurement; typically, a peak observed in a spectrum.

Los Alamos National Laboratory (LANL): National laboratory in northern New Mexico operated by the University of California for the Department of Energy.

macromolecule: Molecule of high relative molecular mass the structure of which usually consists of multiply repeated units that are derived—actually or conceptually—from molecules of low relative molecular mass; particularly a molecule of this kind that is of biological origin.

magnetic field: Modification of free space or vacuum caused by the presence of moving charges that results in a force being exerted on other moving charges.

Magnetic fields are caused by electrical currents, which may be either microscopic (i.e., orbital motion in atoms) or macroscopic, as in the case of an electromagnet.

magnetic moment: Property of a magnetic dipole that determines the amount of torque exerted on it when it is placed in a magnetic field.

magnetic order: Systematic arrangement of magnetic moments in a material that forms a long-range pattern.

magnetic resonance imaging (MRI): Noninvasive technique based on nuclear magnetic resonance for imaging the interior of objects that is often used medically. The sample to be imaged is placed in a strong magnetic field that varies across its volume in a known manner and is then exposed to electromagnetic radiation of appropriate frequency. In this environment the frequency of the NMR signals generated by all the magnetically active atoms in the sample will vary with location in the sample. The 3D distributions of molecules of a particular type in a sample can be reconstructed from the frequency distributions.

magnetism: The attractive and repulsive forces magnets exert on each other. Commonly taken as synonymous with ferromagnetism—that is, the intrinsic magnetic fields characteristic of ferromagnetic materials. More generally, magnetism spans the whole range of phenomena displayed by materials with constituents having magnetic dipoles, including antiferromagnetism and other forms of short-range permanent or ephemeral order.

magnetoresistance: In some materials, electrical resistance depends dramatically on external magnetic field. Antiferromagnetically coupled magnetic layers separated by nonmagnetic spacers can display an extreme form of magnetoresistance called giant magnetoresistance, which is taken advantage of in the magnetic sensors used in high-density disk drives.

MgB_2: Magnesium diboride is a superconductor that has conventional superconducting properties despite having two types of electrons that participate in its superconductivity. Its critical temperature (about 39 K) is the highest of all known phonon-mediated superconductors. This relatively inexpensive material was first synthesized in 1953, but its superconducting properties were not discovered until 2001.

National High Magnetic Field Laboratory (NHMFL): National laboratory for the production of high and specialized magnetic fields for scientific research. It is operated by the National Science Foundation. Its steady-state magnetic field facility is located in Tallahassee, Florida; its pulsed-field facility is based at LANL in New Mexico; and it has an MRI facility and a high field-to-temperature ratio experimental facility at Gainesville, Florida. The NHMFL develops and operates high magnetic field facilities that scientists use for research in physics, biology, bioengineering, chemistry, geochemistry, biochemistry, materials science, and engineering. It is the only facility of its kind in the United States and one of about a dozen in the world.

Nb_3Sn: Niobium-tin (T_c of about 18 K) is a superconducting compound that has been widely used for the construction of high-field magnets with field greater than 10 T or so.

Nb-Ti: Niobium-titanium (T_c of about 9 K) is the workhorse superconducting material in the high-field magnet industry.

neutron source: A nuclear reactor or an accelerator-based facility that generates beams of neutrons of (usually) modest energy that have high intensity and flux. Neutrons are important probes of the microstructure of matter because (1) having no net charge, they interact with the nuclei of atoms, not their electron clouds; (2) they interact weakly with matter and therefore can probe deeply into the interior of samples; and (3) their energies are well suited to the scales of electronic and atomic processes. Moreover, because they are spin-1/2 particles having a magnetic moment, they can be used to study the magnetic microstructure of matter.

NMR spectrometer: Instrument used to measure the frequencies of NMR transitions. A modern NMR spectrometer usually includes (1) a superconducting magnet; (2) a probe for holding the sample in the magnet that includes coils for irradiating it with electromagnetic radiation and detecting the electromagnetic radiation emitted by the sample; and (3) a console, which contains the electronics necessary to operate the probe and a computer to control what happens in the probe and analyze the data returned from the probe.

nuclear magnetic resonance (NMR): When an atomic nucleus in a magnetic field is exposed to photons that have an energy corresponding to the difference in energy between two possible orientations of its magnetic moment, it will resonate—that is, its magnetic moment will rapidly change orientation, in the process first absorbing energy and then radiating it. Only a finite number of different orientations are possible for the magnetic moments of any such nucleus in a magnetic field, each orientation having its own characteristic energy. This behavior is efficient enough that it can be detected over only a narrow range of photon energies (frequencies). The frequencies at which resonances are seen in some specified magnetic field not only identify the kinds of atom responsible for them but can also provide valuable information about the molecular environment in which the atoms are found.

organic superconductors: Class of organic conductors that superconduct at low temperature. They include molecular salts, polymers, and even pure carbon systems—for example, carbon nanotubes and C_{60} compounds. They are also sometimes called molecular superconductors. They are typically large, carbon-based molecules of 20 or more atoms and consist of a planar organic molecule and a nonorganic anion.

pulsed magnet: Resistive magnet designed to provide transient magnetic fields, often for durations as short as microseconds but occasionally for as long as several seconds. Because it is active for only short times, a pulsed magnet uses less power and needs less cooling than a DC magnet of similar bore and maximum field strength. Today, research magnets with the highest fields are pulsed magnets.

quantum critical point: Phase transitions of any sort that occur at absolute zero; thought to be a characteristic feature of all strongly correlated electron systems. The novel behaviors that are observed signal the dominance of quantum fluctuations over the thermal fluctuations that are characteristic of phase transitions at finite temperatures. Many believe that the unconventional properties of high-temperature superconductors may be related to a hidden quantum critical point in these materials.

quantum Hall effect (QHE): When a magnetic field is applied perpendicularly to a thin metal film or a semiconductor film that is conducting an electric current, a voltage will be observed that is perpendicular to the axis of both the film and the magnetic field. This voltage is proportional to the strength of the applied field. Discovered in 1879, this phenomenon was named for its discoverer, Edwin H. Hall. It was later observed that certain superconducting devices display steps in their Hall resistance—that is, the ratio of Hall voltage to current—that reflect the tuning of charge carrier occupancy states by the external magnetic field. K. Von Klitzing was awarded the Nobel prize in physics in 1985 for his demonstration of this phenomenon, which is known as the integer quantum Hall effect.

quantum Hall effect, fractional (FQHE): Fractional version of QHE, in which the Hall resistance progresses in fractions of integer quanta, was discovered in 1982 by

D. Tsui and H. Störmer in experiments performed on gallium arsenide heterostructures. This behavior was explained by R. Laughlin in 1983 in terms of a novel quantum liquid phase that accounts for the effects of interactions between electrons. The three were awarded the 1998 Nobel prize in physics for this work.

quench: The transition, often sudden, in a superconducting material from its superconducting state to its normal, resistive conducting state. It occurs when either the critical current density (J_c) of the material or its critical temperature (T_c) is exceeded.

resistive magnet: Electromagnet that generates a magnetic field by the passage of electric current through resistive conductors.

resistivity: Property of a material that inhibits the flow of electricity, usually because of collisions between the charge carriers and the material's internal lattice structure.

Spallation Neutron Source (SNS): Large research facility under construction at the Oak Ridge National Laboratory. It is expected to be the most powerful pulsed neutron source in the world when completed. The neutrons produced by a spallation source are knocked out of a target (spalled), which is usually a mass of some high atomic weight metal, by high-energy protons generated by an accelerator of some kind.

solenoid: Magnetic solenoid; the most common type of magnet, formed by wrapping coils of conductor around a central cylindrical volume.

spectroscopy: (Usually) the experimental study of the energy levels of materials. More generally, a spectrum is a display of the dependence of some property of a sample as a function of some other parameter—for example, energy absorption versus energy or abundance versus molecular mass. Any experimental activity that generates such plots can be described as spectroscopy.

stored energy: Potential energy; energy that can be released to do work, as in an electric motor. A magnet's energy is stored in its magnetic field.

superconducting magnet: Electromagnet whose conductor is made of superconducting material.

superconductivity: Phenomenon that occurs in certain materials at low temperatures. It is characterized by the complete loss of electrical resistance and the complete expulsion of externally applied (weak) magnetic fields (the Meissner effect).

superconductor: Any material that will conduct electricity without resistance.

superconductor, high-temperature (HTS): Superconducting material that has a high critical temperature. There is no specific temperature separating HTS from LTS materials. HTS now normally also means a CuO_2-based superconductor.

superconductor, low-temperature (LTS): Superconducting materials whose T_c is below about 30 K, though many now call MgB_2 a low-temperature superconductor even though its T_c can be as high as 40 K. See *HTS*.

synchrotron light source: Charged particles traveling in circular trajectories emit electromagnetic radiation. This phenomenon results in a serious loss of ion energy from the circular accelerators used by the high-energy physics community, and some of the radiation emitted can be x rays. A synchrotron light source is an accelerator designed and operated for the electromagnetic radiation it produces rather than for beams of high-energy ions.

T_c: Scientific notation for the critical transition temperature (at zero applied magnetic field and current) below which a material begins to superconduct.

tesla: Unit of measure for magnetic field strength in the SI system of units, abbreviated as T. One tesla is equivalent to 10,000 gauss.

transition temperature: See T_c.

YBCO: Acronym for a well-known ceramic superconductor composed of yttrium, barium, copper, and oxygen.

D

Meeting Agendas

FIRST MEETING
WASHINGTON, D.C.

Thursday, September 4, 2003

Closed Session

8:30 am Welcome
8:45 Introduction to the National Academies and the study process
—Maureen Mellody, Program Officer
9:00 Goals and opening thoughts
—Peter Moore, Chair
10:30 Composition and balance discussion
—Don Shapero, Director
11:30 Discussion of the task and scope of the study

Open Session

1:00 pm Perspectives from the Division of Materials Research at NSF
—Hugh van Horn, Program Director, National Science Foundation

1:30 Perspectives from the Office of Basic Energy Sciences at DOE
—William Oosterhuis, Program Manager, Department of Energy

2:00 Perspectives from the Office of Fusion Energy Sciences at DOE
—Joseph Minervini, Massachusetts Institute of Technology

2:30 Perspectives from the National Institute of Standards and Technology
—J. Michael Rowe, Director, NIST Center for Neutron Research

3:00 Break

3:15 Outcomes of the 1988 Large Magnetic Fields report for NSF
—Frederick Seitz, Rockefeller University

4:30 Perspectives from the commercial sector
—Michael Cuthbert, Oxford Instruments

5:30 Adjourn for the day

Friday, September 5, 2003

Open Session

8:30 am Biology and nuclear magnetic resonance
—Rob Tycko, National Institutes of Health

9:00 Semiconductors and heterostructures
—Mansour Shayegan, Princeton University

9:30 Technology and instrumentation
—Greg Boebinger, NHMFL

10:30 Break

11:00 High-temperature superconductors
—David Larbalestier, University of Wisconsin at Madison

11:30 Magnetic materials
—Meigan Aronson, University of Michigan

Noon International perspectives
—Gabriel Aeppli, University College London

12:30 pm Lunch

Closed Session

1:30 Committee discussions

3:00 Adjourn

SECOND MEETING
TALLAHASSEE, FLORIDA

Monday, December 8, 2003

Open Session

8:30 am	Welcome and goals for the meeting —Peter Moore, Chair
9:00	NHMFL facilities and plans —Greg Boebinger, NHMFL
10:00	Ion cyclotron resonance —Alan Marshall, Florida State University
10:30	Break
11:00	Magnets and high energy physics —Steve Gourlay, Lawrence Berkeley National Laboratory
Noon	Lunch
1:00 pm	Magnetic resonance imaging —Tom Mareci, University of Florida
2:00	Commercial magnet technology —Razvan Teodorescu, Bruker Biospin Corporation
3:00	Break

Closed Session

3:30	Committee discussions
5:30	Adjourn for the day

Tuesday, December 9, 2003

Open Session

8:30 am	Tour of the NHMFL facilities
10:30	Break
11:00	Low-dimensional electron systems —Horst Stormer, Columbia University
Noon	Lunch

Closed Session

1:00 pm	Committee discussions
5:00	Adjourn

THIRD MEETING
WASHINGTON, D.C.

Saturday, March 27, 2004

Closed Session

9:00 am	Discuss conclusions and recommendations
12:15 pm	Working lunch
1:00	Writing group breakout sessions
3:00	Reconvene, discuss report
5:30	Adjourn

Sunday, March 28, 2004

Closed Session

9:00 am	Discuss report findings and recommendations
10:45	Discuss plans for closure and review charge to the committee
Noon	Working lunch
1:00 pm	Adjourn

E

Input from the Community

A broad call for community input to the committee was issued in autumn 2003, shortly after the committee's first meeting. The announcement was sent to several professional societies and appeared on the committee's public Web page. It is excerpted below.

> Dear Colleague,
>
> The National Research Council (NRC) has established a committee called the Committee on Opportunities in High Magnetic Field Science (COHMAG). Its mission is to produce a report on the facilities available to scientists worldwide for doing experiments at high magnetic fields (i.e., at fields above 12 T), the current state of the many scientific disciplines that use high field magnets, the scientific opportunities these fields present, and the prospects for advances in related technologies. With this message COHMAG invites you to send it any information or opinions you feel should be taken into account during its deliberations. Specifically, how have high magnetic fields had an impact on your research? How have you taken advantage of facilities at the National High Magnetic Field Laboratory (NHMFL) or other high-field magnet centers? What new facilities or new capabilities would be most valuable to you? In what new areas of research are high magnetic fields likely to have a large impact? Any other comments?
>
> Why did the NRC set up COHMAG? Earlier this year, the National Science Foundation (NSF) commissioned the NRC to generate a report on the scientific issues that surround the generation of high magnetic fields and their use in scientific research.

Given that the last major report covering this area was issued a decade and a half ago, a new study seems both appropriate and timely.

COHMAG is distributing this message to as many members of the high magnetic field community as possible because it wants to be sure that all voices have been heard before it issues its report. In order to reach as many people as possible, this message is being distributed using e-mail lists obtained from several different organizations, and they, inevitably, are overlapping. We apologize if you have received multiple copies of this message.

If you have information you want to transmit to COHMAG, please communicate it by e-mail to cohmag@nas.edu, and thank you for your help.

For COHMAG,

Peter B. Moore, Chair

Written responses were received from the following individuals:

Richard Beger, National Center for Toxicological Research
Oscar Bernhal, University of California at Los Angeles
Paul Canfield, Ames Research Center
Walter Chazin, Vanderbilt University
David Cowburn, New York Structural Biology Center
Jack Crow, NHMFL
Kwaku Dayie, Cleveland Clinic Foundation
M. Dolotenko, Russian Federal Nuclear Center
Thomas Erber, Illinois Institute of Technology
Bolzonia Fulvio, IMEM, Italy
Roy Goodrich, Louisiana State University
Jurgen Haase, Leibniz Institute, Dresden
Michael Hall, Texas A&M University
William Halperin, Northwestern University
Bruce Hammer, University of Minnesota
Fritz Herlach, Katholieke Universiteit Leuven
Seung Hong, Oxford Instruments
Robert Leif, Newport Instruments
Gerard Ludtka, Oak Ridge National Laboratory
Gerhard Martinez, Grenoble High Magnetic Field Lab
Andrew Maverick, Louisiana State University
Craig Milling, Magnetic Resonance Microsensors
Martha Morton, University of Connecticut
William Moulton, NHMFL
Jan Musfeldt, University of Tennessee

Florin Neascu, International School of Choeifat, United Arab Emirates
Dean Peterson, Los Alamos National Laboratory
Neela Poorasingh, City University of New York
Al Redfield, Brandeis University
Jim Rhyne, Los Alamos National Laboratory
Larry Rubin, Massachusetts Institute of Technology
Joshua Telser, Roosevelt University
Cees Thieme, American Superconductor Corp.
Sheldon Schultz, University of California at San Diego
Horst Stormer, Columbia University
David Weber, University of Maryland
Roy Weinstein, University of Houston
Nicholas Zumbulyadis, Eastman Kodak

F

Biographies of Committee Members and Staff

COMMITTEE MEMBERS

Peter B. Moore, ***Chair,*** is Sterling Professor of Chemistry at Yale University, where he is also a faculty member in the Department of Molecular Biophysics and Biochemistry. He received his Ph.D. in Biophysics from Harvard University in 1966. He joined the Yale faculty in 1969 following postdoctoral fellowships at the Institute de Biologie Moleculaire, Geneva, Switzerland, and at the MRC Laboratory of Molecular Biology, Cambridge, U.K.. The focus of Dr. Moore's research is the delineation of the relationship between structure and function in RNAs and ribonucleoproteins. He is best known for his work on the ribosome, the enzyme that catalyzes mRNA-directed protein synthesis, and prominent among the experimental techniques he has used are neutron scattering, NMR, and x-ray crystallography. He has been a member of numerous editorial boards and advisory committees, having recently completed a 5-year term as the editor of *Biophysical Journal.* Dr. Moore was elected to the National Academy of Sciences in 1997 and to the American Academy of Arts and Sciences in 2003. He is a fellow of the AAAS and the Biophysical Society.

Gabriel Aeppli is the Quain Professor of Physics at the Department of Physics and Astronomy of University College London (UCL). He is also the director of the London Center for Nanotechnology, a joint enterprise between Imperial College London and University College London. Prior to taking up his UCL position,

Dr. Aeppli was a senior research scientist with the NEC Research Institute in Princeton and a former Distinguished Member of the Technical Staff at AT&T Bell Laboratories. He received his Ph.D. from the Massachusetts Institute of Technology (MIT) in 1983. His research interests are in the application of neutron scattering techniques to the investigation of magnetism, superconductivity, and micromagnetism. More recently he has turned his attention to quantum information processing and medical diagnostics, both of which exploit micro- and nanotechnology. He is a fellow of the American Physical Society, was a corecipient of the 2003 International Magnetism/Neel prize, and serves on numerous national and international review committees, including the DOE Basic Energy Sciences Advisory Committee's Subcommittee on Pulsed Spallation Source Upgrades. Dr. Aeppli has pioneered accelerator- and reactor-based neutron scattering techniques to measure magnetic excitations in solids. He was a member of the NRC committee the wrote the decadal survey of condensed matter and materials physics and served on the Solid State Sciences Committee.

Meigan Aronson is a professor of physics and associate dean for natural sciences at the University of Michigan. She is also asssociate director of the Michigan Electron Microbeam Analysis Laboratory, a user facility for the university research community. Dr. Aronson earned a Ph.D. from the University of Illinois, Urbana-Champaign, in 1988. Her research is on quantum phase transitions, phase behaviors of low-density metals, and novel magnetism. Her research focuses on the exploration of magnetism in metals and the properties of the electron gas at low densities, where strong and unscreened Coulomb interactions are expected to lead to unusual types of charge and spin order, especially in very large magnetic fields. Her group uses neutron scattering as well as a variety of transport, magnetic, and thermal measurements to probe the ground state and its excitations at low temperatures, high magnetic fields (up to 60 T), and pressures as great as 200,000 atm. Dr. Aronson is a fellow of the American Physical Society.

Paul M. Chaikin is the Henry DeWolf Smyth Professor of Physics at Princeton University and a faculty member at the Princeton Materials Institute. Dr. Chaikin, who received his Ph.D. in physics from the University of Pennsylvania, is coauthor of *Principles of Condensed Matter Physics*, a definitive book on this subject. His experimental investigations in hard condensed matter (quantum electronic physics and low-temperature physics) and soft condensed matter (statistical mechanics of phase transitions, colloids, polymers, hydrodynamics) and at the interface often use physics and techniques from one subfield in the other. He is particularly interested in the effects of dimensionality, Coulomb correlation, and disorder in condensed-matter systems, spin-density-wave states and superconductivity in organic metals, superconducting wire networks, colloidal physics, and sedimentation in fluidized beds. Dr. Chaikin is a fellow of the American Academy of Arts and Sciences, a fellow of the American Physical Society, a member of the National Academy of Sciences, and a past winner of the Guggenheim Fellowship and the A.P. Sloan Foundation Fellowship.

Paul D. Ellis is a laboratory fellow at Pacific Northwest National Lab (PNNL) and chief scientist of the William R. Wiley Environmental Molecular Sciences Laboratory's high-field magnetic resonance facility. In 1970, he received his Ph.D. in chemistry from the University of California at Davis, where he had also studied for his B.S. Dr. Ellis spent 23 years at the University of South Carolina as a professor of chemistry, where he was selected as the George H. Bunch, Sr., Professor of Science in 1984. In 1993 he left the university to accept the position of associate director of the Macromolecular Structure & Dynamics Directorate at PNNL. Dr. Ellis's expertise is in three main fields: nuclear magnetic resonance spectroscopy, structural biology, and heterogeneous catalysis. Among his awards and committee memberships are these: Russell Award for Science and Engineering (1976), A.P. Sloan Foundation Fellow (1977-1979), editorial advisory board for *Magnetic Resonance in Chemistry* (1984), South Carolina Chemist of the Year (1985), advisory committee for the Biological Facilities Centers Program, NSF (1987-1989), editorial advisory board for *Concepts in Magnetic Resonance* (1989), corecipient of the South Carolina Governor's Award for Excellence in Science (1990), World Bank Visiting Scholar in China (1991), member of various study sections at the National Institutes of Health (NIH)—General Medical Sciences (1991-1995) and NIH Reviewers Reserve (1995-1999)—and fellow of the American Association for the Advancement of Science (2003). Dr. Ellis is also a member of the Natural Sciences and Engineering Research Council (NSERC) of Canada's Major Facilities Access Committee (2004).

Peter F. Green holds the B.F. Goodrich endowed professorship in materials engineering at the University of Texas at Austin, where he is currently a professor of chemical engineering. Dr. Green is a fellow of the American Physical Society (1996) and of the American Ceramic Society (1998). He is a divisional associate editor for *Physical Review Letters* and serves on the editorial boards for *Macromolecules* and the *Journal of Polymer Science: Polymer Physics.* He currently serves on the external advisory committee of the Division of Math and Physical Sciences of the National Science Foundation. Additionally, Dr. Green is a member of the Council of Gordon Research Conferences and is serving as vice president for the Materials Research Society. Dr. Green received his Ph.D. degree in materials science from Cornell University in 1985. Following postdoctoral research in the Ions, Solids, and Interactions Department at Sandia National Laboratories, he joined the staff of the Physical Properties of Polymers Division in 1986. From 1991 to 1996 he served as Department Manager of the Glass and Electronic Ceramics department before moving to the University of Texas at Austin in 1996. Dr. Green is presently vice chair of the NRC's Solid State Sciences Committee. His research interests broadly encompass problems associated with the structure and dynamics of oxide glass melts and polymeric melts, wetting, and interfacial phenomena in soft materials.

David C. Larbalestier is a professor in the Department of Materials Science and Engineering and in the Department of Physics at the University of Wisconsin, Madison, where he holds both the L.V. Shubnikov Chair and the Grainger Chair of Superconductivity. He is also the director of the Applied Superconductivity Center, an interdisciplinary center of about 40 faculty, staff, and students. He has been active in superconductivity since his Ph.D., for which his dissertation gained the Matthey Prize of Imperial College. After 2 years in Switzerland, he returned to England in 1972 to the Superconducting Magnet Research Group of the Rutherford Laboratory, working for 4 years on the development of multifilamentary Nb_3Sn conductors and magnets. This work culminated in the first filamentary Nb_3Sn NMR magnet, for which he shared a 1978 industrial research award (IR-100) with an Oxford Instrument Company team. He joined the University of Wisconsin in 1976, becoming associate chairman of his department in 1981, a post held until 1991, when he became director of the Applied Superconductivity Center. Dr. Larbalestier has been exceptionally active in promoting collaborations uniting industry, national laboratories, and other university groups, asserting a leadership role in both the low-temperature and high-temperature superconducting materials communities. He has been recognized by prizes of the IEEE (1991 and 2000) and the Council for Chemical Research (2000) for his work and that of his collaborators on $(Bi,Pb)_2Sr_2Ca_2Cu_3O_x$. He has served on many review panels of the National Science Foundation and the Department of Energy, was a member of the 1987 National Academy of Sciences Panel on High Temperature Superconductivity, and led the 1996 World Technology Evaluation Center Panel on Energy Applications of Superconductors, sponsored by DOE and NSF. He has published more than 250 refereed papers and given more than 200 seminars and presentations at scientific meetings. During 2000 he was Visiting Professor at the University of Geneva and Visiting EPSRC Fellow at the Imperial College of the University of London. He was elected to the National Academy of Engineering in 2003. Born a U.K. subject, Dr. Larbalestier became a U.S. citizen in 1988.

J. David Litster is a professor of physics at the Massachusetts Institute of Technology. Dr. Litster received a bachelor's degree in engineering from McMaster University in Hamilton, Ontario, and a Ph.D. in physics from MIT. He was appointed to the faculty at MIT in 1966 and became a professor of physics in 1975. Dr. Litster's research interests have been the experimental study of phase transitions in unusual states of matter, using primarily light-scattering and high-resolution x-ray scattering. He is a former chair of the NRC's Solid State Sciences Committee (1993 1995). From 1988 to 1991, Dr. Litster was director of the MIT Francis Bitter National Magnet Laboratory. For 5 years before, he was director of the MIT Center for Materials Science and Engineering. From 1979 to 1983, he headed the Division of Condensed Matter, Atomic and Plasma Physics in the MIT Department of Physics. From 1991 to 2001, Dr. Litster served as vice president and dean of research. From 1996 to 2001, he also served as dean for graduate education. Dr. Litster is a fellow of the American Physical Society, the American Academy of Arts and Sciences, and the American Association for the Advancement of Science. He received an honorary doctorate from McMaster University in 1992 and the Irving Langmuir Prize in Chemical Physics from the American Physical Society in 1993.

Joseph Minervini is division head for technology and engineering in the Plasma Science and Fusion Center at the Massachusetts Institute of Technology. He also holds an academic appointment as senior research engineer in the Nuclear Engineering Department, where he teaches a course and supervises graduate student research. He is spokesperson for the U.S. Magnetics Program organized under the Virtual Laboratory for Technology of the DOE Office of Fusion Energy Science (OFES). Dr. Minervini's research interests include applied superconductivity, electromagnetics, cryogenic heat transfer, supercritical helium fluid dynamics, and low-temperature measurements. He has worked on magnet systems for nearly every major application of large-scale superconductivity, including fusion energy, magnetic levitation, energy storage, power generation, magnetic separation, high-energy physics, and medical applications. He has published over 70 papers in these fields. His most recent large project was as principal investigator for the magnetics R&D program supported by OFES as part of the International Thermonuclear Experimental Reactor (ITER) project. Dr. Minervini holds a B.S. in engineering from the U.S. Merchant Marine Academy, Kings Point (1970), and S.M. (1974) and Ph.D. (1981) degrees in mechanical engineering from the Massachusetts Institute of Technology. He also serves as an international advisory editor for *Cryogenics* and served on the editorial advisory board for *Fusion Engineering and Design*. He is a board member of the Applied Superconductivity Conference and serves on the board of the Cryogenic Engineering Conference and is a member of the standing committee for the Symposium on Fusion Engineering. Dr. Minervini is a member of the American Society of Mechanical Engineers.

J. Michael Rowe retired from the National Institute of Standards and Technology in 2003. Dr. Rowe received a B.Sc. in engineering physics from Queens University in Kingston, Ontario, and a Ph.D. in physics from McMaster University. In 1966 he joined Argonne National Laboratory, and in 1973 he joined the staff of the National Institute of Standards and Technology (NIST, formerly NBS). His research interests are in the area of neutron scattering from condensed matter, with a particular focus on simple liquids, hydrogen in metals, and phase transitions in molecular crystals. In 1986 he became the manager of the Cold Neutron Project, which developed the only fully competitive capability in the United States for research using cold neutrons. He served as the director of the NIST Center for Neutron Research from 1989 until his retirement. He is a fellow of the American Physical Society and the American Association for the Advancement of Science. Dr. Rowe received Presidential rank awards as a Distinguished Federal Executive in 1992 and Meritorious Federal Executive in 2003, and received the Samuel Wesley Stratton Award in 1994 for his research on hydrogen in metals. In 2004, he was the first recipient of the Clifford G. Shull Prize in Neutron Science given by the Neutron Scattering Society of America. He has served on many advisory panels and committees, including the Solid State Sciences Committee and Basic Energy Science Advisory Committee.

John M. Rowell holds a Distinguished Visiting Professor position at Arizona State University. He graduated from Oxford University (B.S., M.S., Ph.D.). He worked at Bell Telephone Laboratories for 23 years as a scientist, department head, and director. At Bellcore for 5 years he was assistant vice president of solid state science and technology. He then joined Conductus, a start-up company formed after the discovery of high-temperature superconductivity, where he worked for 6 years as chief technical officer. Dr. Rowell is renowned for his first observation, with P.W. Anderson, of the Josephson Effect and holds the first patent granted for an electronics application of that phenomenon. With W.L. McMillan he developed tunneling spectroscopy. With M. Gurvitch and a number of Bell Labs colleagues he invented the niobium/aluminum trilayer process, which is now used universally for SQUIDs and for digital circuit fabrication. He was awarded the Fritz London Memorial Low Temperature Physics Prize with W.L. McMillan, is a fellow of the Royal Society, and is a member of the National Academy of Sciences and the National Academy of Engineering. He has served on a number of NRC committees, including the Panel on Superconductivity (1987), the Panel on Condensed-Matter Physics (1983-1986), and the Panel on Artificially Structured Materials (1984-1985).

Mansour Shayegan is a professor in the Department of Electrical Engineering at Princeton University. He received his B.S. (1979), M.S. and E.E. (1981), and Ph.D.

(1983) degrees in electrical engineering from the Massachusetts Institute of Technology. He spent 2 years as a research associate in the Department of Physics and Astronomy at the University of Maryland before joining the faculty at Princeton in 1985. His research activities include semiconductor physics, electronic properties of low-dimensional semiconductor structures, and growth of III-V compounds by molecular beam epitaxy. In 1986, he received an NSF Presidential Young Investigator Award and an IBM Faculty Development Award, in 1989 an Alfred P. Sloan Fellowship, in 1993 a Fulbright Fellowship for lecturing and research in Germany, and in 1997 an Alexander von Humboldt Fellowship for research in Germany. Dr. Shayegan is a fellow of the American Physical Society.

Robert Tycko is the chief of the Solid-State NMR and Biomolecular Physics Section of the Laboratory of Chemical Physics in the National Institute of Diabetes and Digestive and Kidney Diseases at the National Institutes of Health. He has held this position since 1998. Dr. Tycko received his Ph.D. from the University of California at Berkeley in 1984. Prior to joining the NIH in 1994, Dr. Tycko spent 8 years as a member of the technical staff in the Materials Chemistry Research Department of AT&T Bell Laboratories. His research focuses on the development of solid-state NMR methods for structural studies of biopolymers and on the application of these methods to problems in biophysics, biochemistry, and structural biology. Examples of systems of interest include peptides and proteins that interact with biological membranes, amyloid fibrils associated with Alzheimer's disease, and disordered states of proteins involved in protein folding processes. Dr. Tycko's research is highly interdisciplinary, drawing heavily on experimental and theoretical techniques of both the physical and the biological sciences. He is a fellow of the American Physical Society (1997) and has received the NIH Director's Award (2001), the Max T. Rogers Distinguished Lectureship (2003), and the R.W. Vaughan Lectureship (2004). Dr. Tycko is currently a member of the editorial boards of the *Journal of Magnetic Resonance* and the *Journal of Chemical Physics.* He served as chair of the Gordon Research Conference on Magnetic Resonance in 2001.

Valerii Vinokur is a senior scientist at Argonne National Laboratory, where he is also director of the Materials Theory Institute in the Materials Science Division. His research interests include the dynamics of disordered media and nonequilibrium stochastic dynamics; disordered superconductors: flux pinning, flux dynamics, and related phenomena; mesoscopic superconductivity and magnetism, spintronics, and coupled magnetosuperconducting nanostructures; and electronic transport in quantum and low-dimensional systems. Prior to joining Argonne in 1990, Dr. Vinokur held appointments at the Eidgenössische Technische Hochschule (ETH) Zürich Institut für Theoretische Physik (ITP) in Zürich, Switzerland, Leiden

University in the Netherlands, and the USSR Academy of Science. He received his Ph.D. in 1979 from the USSR Academy of Science Institute of Solid State Physics. He has been a Fellow of the American Physical Society since 1998, and received the 2003 John Bardeen Prize for Theory of Superconductivity for his work on the Theory of Vortex Matter.

NRC STAFF

Donald C. Shapero, director of the Board on Physics and Astronomy, received a B.S. degree from the Massachusetts Institute of Technology in 1964 and a Ph.D. from there in 1970. His thesis addressed the asymptotic behavior of relativistic quantum field theories. After receiving the Ph.D., he became a Thomas J. Watson postdoctoral fellow at IBM. He subsequently became an assistant professor at American University, later moving to Catholic University and then joining the staff of the National Research Council in 1975. He took a leave of absence from the NRC in 1978 to serve as the first executive director of the Energy Research Advisory Board at the Department of Energy. He returned to the NRC in 1979 to serve as special assistant to the president of the National Academy of Sciences. In 1982, he started the NRC's Board on Physics and Astronomy (BPA). As BPA director, he has played a key role in many NRC studies, including the two most recent surveys of physics and the two most recent surveys of astronomy and astrophysics. He is a member of the American Physical Society, the American Astronomical Society, and the International Astronomical Union. He has published research articles in refereed journals in high-energy physics, condensed-matter physics, and environmental science.

Timothy I. Meyer is a program officer at the NRC's Board on Physics and Astronomy. He received a Notable Achievement Award from the NRC's Division on Engineering and Physical Sciences in 2003 and a Distinguished Service Award from the National Academies in 2004. Dr. Meyer joined the NRC staff in 2002 after earning his Ph.D. in experimental particle physics from Stanford University. His doctoral thesis concerned the time evolution of the B meson in the BaBar experiment at the Stanford Linear Accelerator Center. His work also focused on radiation monitoring and protection of silicon-based particle detectors. During his time at Stanford, Dr. Meyer received both the Paul Kirkpatrick Award and the Centennial Teaching Award for his work as an instructor of undergraduates. He is a member of the American Physical Society, the American Association for the Advancement of Science, and Phi Beta Kappa.

G

Tutorial on High-Temperature Superconductivity

High-temperature superconductor materials and low-temperature superconductor materials differ markedly in charge response, as measured by transport and optical experiments; spin response, as indicated by static susceptibility, NMR, and inelastic neutron scattering experiments; and in single-particle spectral density, as evidenced by angle-resolved photoemission studies (ARPES).

The basic observations that appear to be crucial for understanding the mechanism of high-temperature superconductivity can be summarized as follows.

- The action occurs primarily in the Cu-O planes, so that it suffices, in first approximation, to focus both experimental and theoretical attention on the behavior of their planar excitations and to focus as well on the two best-studied systems, the 1-2-3 system ($YBa_2Cu_3O_{7-x}$) and the 2-1-4 system ($La_{2-x}Sr_xCuO_4$).
- At zero doping and low temperature, both of the systems ($YBa_2Cu_3O_{7-x}$ and $La_{2-x}Sr_xCuO_4$) are antiferromagnetic (AF) insulators with an array of localized Cu^{2+} spins that alternate in sign throughout the lattice.
- Holes may be injected into the Cu-O planes of the 1-2-3 system by adding oxygen; for the 2-1-4 system, strontium can be added. The resulting holes on the planar oxygen sites bond with the nearby Cu^{2+} spins, making it possible for the other Cu^{2+} spins to move and, in the process, destroying the long-range AF correlations found in the insulator.

- If one adds sufficient holes, the system changes its ground state from an insulator to a superconductor.
- In the normal state of these superconducting materials, the itinerant, but nearly localized, Cu^{2+} spins form an unconventional Fermi liquid, with its quasi-particle spins displaying strong AF correlations even for systems at doping levels greater than those at which T_c is maximum—so-called "overdoped" materials.

Measurements of HTS electronic properties in magnetic fields will be the key component of the experimental work that must be done before a full understanding of the HTS phenomenon will be attained. The main issues that can be addressed this way include these:

- Effects dependent on the quantization of electron states in high magnetic fields and so-called quantum oscillations (QO), of which two kinds can be distinguished: (1) oscillations of magnetic susceptibility (de Haas-van Alphen effect (dHvA)) and (2) oscillations of conductivity (Shubnikov-de Haas effect) and
- Cyclotron resonance (CR).

For CR effects to be observed, $\omega_c\tau >> 1$, where ω_c is the cyclotron frequency and τ is the electron relaxation time. The maximum value of τ at T_c in optimally doped YBaCuO ($T_c \approx 92$ K) is 10^{-13} s, and it drops by factor of 2-3 at room temperature. In order to observe CR or QO, the material must be in the normal state; for a typical ω_c of $\sim 10^{14}\,s^{-1}$, one needs $T > 90$ K and $H > 500$ T. Even higher fields may be necessary if the effective mass of the carriers in the material is greater than that of a free electron, and this explains why preliminary dHvA measurements made on optimally doped YBaCuO at $H < 250$ T did not give conclusive results.

Superfluid density, ρ_s, in HTS materials is a key quantity that speaks to the nature of high-temperature superconductivity, and it can be studied using high magnetic fields. At zero temperature, ρ_s is a direct measure of the electronic states that participate in superconductivity, and at low temperatures, it probes low-lying, quasi-particle states, reflecting the symmetry of the order parameter. At intermediate temperatures, the superfluid density follows the temperature behavior of the order parameter and can reveal normal-state anomalies below T_c. Superfluid densities are obtained from measurements of the London penetration depth. Optical experiments combined with DC transport, AC penetration depth, and tunneling measurements will be instrumental in answering these long-standing questions about the nature of high-temperature superconductivity.

An interesting feature of cuprate superconductors is that despite general condensate formation, not all carriers appear to participate in superconductivity. A considerable fraction of the carriers' energy spectral weight does not condense.[1] In optimally doped HTS materials, only a modest fraction (~1/5) of the spectral weight condenses, and even less condenses in under doped cuprates. In this regard, HTS materials are very different from clean metallic superconductors, in which essentially every carrier pairs up and participates in superconductivity. In general, superfluid density arises from the condensation of the spectral weight from the broad interval of frequencies into a delta function at zero frequency. Thus, the study of superfluid densities in HTS materials may help reveal the nature of this puzzling behavior.

The study of the effects of magnetic fields on superfluid density will thus be one of the central directions of HTS experimental studies in the years to come. Furthermore, the behavior of ρ_s as a function of temperature in high magnetic fields needs to be explored. Penetration depth measurements at frequencies between the kilohertz and terahertz regimes indicate a significant reduction in superfluid density in modest magnetic fields. Optical studies should be undertaken of underdoped as well as overdoped materials. Another important direction is the study of the superconductor-insulator transition at high magnetic fields by observing the shift of the spectral weight from low to high frequencies as a superconducting or normal state appears in the excitation spectrum.

The behavior of ρ_s at zero temperature as a function of doping could also shed new light on the possible non-Fermi liquid nature of HTS materials. In the underdoped region, ρ_s increases with the doping parameter x (as in -O_{7-x}) and is approximately proportional to increases in T_c. In the overdoped region, at first ρ_s increases with x beyond the optimal hole concentration despite the reduction in T_c, and then there is a final and abrupt decrease in ρ_s even if the hole concentration continues to increase.

One can thus think of the following objectives for further research:

- To understand why both YBCO and BSCCO have superfluid densities ρ_s at zero temperature five times smaller than expected from general Bardeen-Cooper-Schrieffer (BCS) considerations, and yet why at the same time the ratio $\rho_s(T)/\rho_s(0)$ agrees with BCS predictions.

[1] In these materials, the carriers' energies are distributed across an energy spectrum. Experiments have shown that carriers of certain energies do not condense and so do not participate in superconductivity.

- To provide conclusive evidence that over-doped cuprates are not classic Fermi liquids, and to resolve the question of whether there is a quantum critical point at zero temperature in the superconducting domain.
- To explore the differences between the electron- and hole-doped materials and whether the distinctions between them in d-wave pairing behavior hold in the superconducting state.

The effect of magnetic fields perpendicular to Cu-O planes is also of great interest. Vortex arrays that form in a superconducting domain govern the magnetic and transport properties (see Appendix H). The quasi particles that appear near the nodes of the *d*-wave gap may make an essential contribution to the dynamic properties of these vortices. Measurements of optical conductivity in magnetic fields under different doping conditions will be critical to uncovering these quasi-particle effects. At the same time parallel (to the CuO planes) field studies could break new ground in cuprates. Indeed, one expects that in very thin films, the Zeeman energy will be sufficient to split low-lying "spin-up" and "spin-down" quasi particles. Thus, a portion of the Fermi surface near the nodes of the energy gap will become normal, even at 0 K, and this effect will be detectable because of the corresponding reduction in $\rho_s(0)$. This state indeed would be a novel superconducting one.

H

Tutorial on Frontiers in Vortex Physics

This appendix introduces the study of vortex matter and describes some of the leading avenues of research. The experimental exploration of vortex matter, which is one of the many topics uncovered by HTS research, has become an important part of modern condensed-matter physics and depends on high magnetic fields.

The study of vortex matter in high fields over the full phase diagram is one of the most appealing research areas in condensed-matter physics. It is interesting not only because vortex system properties govern the charge transport and magnetic properties of HTS materials and, more generally, all Type II superconductors, but also because the behavior of these complex objects is generic for almost all disordered condensed-matter systems, so that cutting-edge contemporary concepts, involved theoretical methods, and the most advanced experimental techniques will be required to understand them. The study of vortices may lead to the physical realization of systems in which the entire statistical and quantum mechanics of strongly correlated systems with tunable parameters can be studied experimentally. A complex interplay between quantum and thermal fluctuations and disorder determines both the thermodynamic and the dynamic properties of vortex matter. Studies of vortex matter melting and the role of disorder have advanced the knowledge of solid and liquid thermodynamics. Last, but not least, several new phenomena, such as dynamic melting, have been discovered in vortex matter. These discoveries have led to the development of a novel and hitherto almost unexplored branch of physics, the physics of dynamic phase transitions. Another example of such a phenomenon is the Josephson plasma resonance effect, which holds great technological promise, as evidenced by the high level of Japanese investment in this area.

A vortex system is a tunable, correlated system. The density of vortex lines in a superconductor depends on the applied magnetic field, so that by varying the external field, one can control the density of the "particles" in the system and hence also their interactions. Vortex separation is ~80 nm at 1 T, 14 nm at 10 T, 6 nm at 50 T, and 4 nm at 100 T. The radius of strong interaction between vortices is determined by λ, generally 200-400 nm for HTS materials. At distances exceeding this scale, interactions are screened and become exponentially weaker. Most experiments—as well as practical applications—are conducted in the range where the vortex spacing is much less than the London depth, which ensures retention of the tunable property of these strongly correlated systems. Additionally, the degree of vortex disorder can often be changed by altering its density using magnetic field strength as the controlling variable. Thus, by adjusting both the temperature and the magnetic field one can scan the entire range of possible interactions.

In general, the quantum mechanics of particles in N dimensions is equivalent to the statistical mechanics of elastic lines in an $N + 1$ dimensional space. Thus the physics of vortex systems in a bulk sample can be mapped onto the quantum mechanics of a two-dimensional system of repulsively interacting bosons. In particular, a vortex lattice can correspond to a two-dimensional Wigner crystal, while a vortex liquid maps onto a two-dimensional superfluid. As tunable systems, vortex states in high magnetic fields offer a unique possibility for study of the fundamental localization problem in strongly correlated systems.

One of the most important questions being investigated today is how the mixed superconducting state depends on the collective properties of the vortices when a transport current is applied. Hydrodynamic interactions of the applied current with the circular currents of the vortices cause the vortices to migrate (via the so-called Magnus force), and energy is dissipated. Thus, in the presence of vortices, a superconductor cannot carry a current without energy loss unless the vortices are pinned in place. Since the first field penetration at H_{c1} occurs at ~0.01 T, even weak currents generate vortices in superconducting devices.

A closely related topic is the behavior of very small superconductor samples in relatively high magnetic fields (a few tesla). For instance, transport properties of superconducting point contacts, which will be an issue in most of the next generation of superconducting devices, is an increasingly important research topic. At the tip of such a contact, superconductivity disappears, and charge transport occurs exclusively via Andreev states. Extensive theoretical research is being carried out on this kind of nanoscale superconductivity, and corresponding experimental investigations are needed. Another important topic concerns superconductor samples where one of the physical dimensions is smaller than the superconducting coherence length, which is a situation often seen in thin superconducting films, wires, and granules. An applied magnetic field induces a quantum phase transition between the superconducting and normal states at moderate (of the order of 1 T) fields in such samples. Measurement of the transport and thermodynamic properties of nanometer-size superconductors in the vicinity of their critical points should result in dramatic advances in the understanding of critical quantum behavior.

Experimental studies of vortex lattice materials can provide insights into phenomena that cannot be as easily accessed with the usual atomic solids. By introducing disorder in vortex lattice materials in a controlled way, using irradiation with high-energy electrons, protons, or heavy ions or chemical means, systems can be obtained in which the thermodynamic properties of glasses can be studied with a precision that is not achievable with other systems, such as spin glasses in random magnets. The study of vortex physics thus offers unique opportunities for exploring the fundamental statistical mechanics of condensed-matter systems.

The superconducting state is characterized by the macroscopic quantum coherence of the whole sample, and it has been experimentally observed that melting of vortex lattices results in a loss of quantum coherence within superconductors. However, it is not known if the lattice melting in question is associated with a complete loss of coherence. There may be a range of magnetic fields and temperatures for which the vortex liquid loses only its transverse coherence, and correlations along the vortex lines persist. Experiments using the Josephson plasma resonance effect in high magnetic fields may provide decisive answers.

There are three main directions in the study of vortex matter today:

- *What factors control the pinning of vortices?* A defining scientific and technological property of superconductors is their ability to carry current without loss. Vortices are present in superconductors in all practical situations, and when a current is applied to a superconductor, the vortices present will start to move unless immobilized by inhomogeneities in the material, which can be either natural or artificially engineered. If the vortices move, there will be energy loss. The effectiveness of this immobilization, or "pinning," effect in any given superconducting material is quantified by its so-called critical current, the maximum current that it can carry without dissipation. Finite critical currents exist only in the vortex solid state and become zero as the vortex lattice melts. The quest to increase the critical currents of existing superconducting materials has thus motivated studies of vortex pinning and research into the thermodynamics of the vortex lattice.
- *How do vortices behave dynamically?* Even at low temperatures, and with small applied currents, vortices slowly move owing to thermal or quantum

fluctuations. This effect is known as vortex creep. Vortex dynamics is the study of pinning-dominated vortex creep.

- *What (other) novel phenomena do Type II superconductors exhibit?* The search for qualitatively new effects that have no analogue in nonvortex matter or that were previously overlooked has become the third important line of investigation. Dynamic phase transitions and the so-called Josephson plasma resonance are good examples of phenomena of this kind.

THERMODYNAMIC PROPERTIES OF VORTEX MATTER

Phase transitions in the presence of disorder are a major concern in condensed-matter physics, and the melting of vortex lattices is a first-order phase transition that has received a lot of attention. In this case, the disorder in question is provided by impurities, crystalline defects, and the like, in the material in which the vortices exist, or it can be artificially introduced into the material via various irradiation methods. Though a good deal is known about the general behavior of this kind of melting, very little is known about the local fluctuations and the factors that determine the local properties of this phase transition. Owing to high anisotropy and very short coherence length (on the order of tens of angstroms), thermal fluctuations play a primary role in determining the physical properties of HTS materials. In conventional superconductors, the vortex-melting line was not detected since it lies very close to the upper critical field H_{c2}. In HTS materials, however, the melted phase—the vortex liquid—occupies a significant part of the vortex phase diagram. Experimentally, vortex melting was unambiguously established by a combination of transport measurements that revealed the onset of finite resistivity marking the disappearance of vortex pinning; neutron scattering measurements that demonstrated a loss of the long-range order; and thermodynamic measurements that showed a peak in heat capacity, which is the unambiguous signature of a thermodynamic phase transition.

The thermodynamics of vortex systems are governed by the interplay between disorder and thermal fluctuations. Disorder transforms the vortex lattice into a vortex glass, a pinning-dominated phase with a highly nonlinear response to small external currents. Artificially manufactured defects can be introduced into a superconductor by irradiating it with heavy ions such as Pb, Au, or U. These impurities enhance pinning because of a geometrical match between the pinning site (the cylindrical amorphous track of the heavy ion through the material) and the linear vortex. Such impurities can extend the vortex glassy phase to higher fields and temperatures, forming a so-called Bose glass. Interestingly, a high density of defects can destroy a topologically defect-free vortex lattice—disorder-induced melting—and lead to the formation of an amorphous vortex glass phase at higher magnetic fields.

The dynamics of nonlinear creep motion is governed by the fact that interactions between disorder and elastic degrees of freedom give rise to effective motion barriers that grow as a power of the applied force. One of the fundamental questions of the physics of glasses is how these growing barriers relate to the proposed hierarchical structure of the glass itself. Studies of noise associated with vortex motion will also advance understanding of the statistics of low-lying states in glasses.

DYNAMIC PROPERTIES OF VORTEX SYSTEMS

Even before the discovery of high-temperature superconductors, it was predicted that the thermally activated dynamics of disorder-dominated systems with internal degrees of freedom would exhibit highly nonlinear responses. For instance, in response to an infinitesimal applied driving force, the system velocity v appeared to show not only a nonlinear but also a nonanalytic dependence on the force F: $v \sim \exp(-\text{const}/F\mu)$, where the exponent, μ, depends on the dimensionality and the characteristic scale of the moving elastic medium and on the dimensionality of the space this medium moves in. For a magnetic domain wall moving in a magnetic film, $\mu = 1/4$; for the vortex lattice in a bulk superconductor at very small currents, $\mu = 1/2$. This theoretical result is highly counterintuitive: Textbook wisdom suggests that all physical systems demonstrate a linear response—that is, v is proportional to F (e.g., Ohm's law) when forces are small. Initially, this result was met with skepticism and disbelief.

Experiments on magnetic relaxation in HTS materials in 1990 confirmed that vortex motion at low temperatures obeys the prediction just discussed. Since then, extensive studies of the peculiarities of vortex dynamics in low temperature vortex phases—vortex and Bose glasses—has become the focus of both experimental and theoretical research on superconductors. Investigation of vortex creep dynamics has been extended to the quantum case (quantum creep), and vortex creep now appears to be generic in all disordered systems. Recently, creep motion was even found for the motion of domain walls and dislocations. It is thus a fundamental characteristic of glassy phases, and the recognition of its importance has motivated intense interest in the dynamics of low-temperature vortex creep, one of the most exciting and fundamental discoveries in the physics of superconductivity of the last decade.

The dissipation processes associated with the vortex motion are related to the scattering of the normal carriers confined within the normal vortex core—the so-called quasi-particle Andreev states. Experimental studies of the AC response of vortex lattice materials in high magnetic fields should provide access to these mechanisms and will help elucidate the microscopic nature of the electronic states in HTS materials.

Another fundamental phenomenon discovered in the context of vortex lattice motion—in this case in the high-driving-force regime—is dynamic melting. When driven in the random environment, the periodic structure (vortex lattice) experiences random forces due to encounters with randomly distributed pinning sites. These collisions cause random fluctuations in the positions of vortices. The effect of this positional disorder therefore resembles the effect of temperature, and one can discuss it in terms of the temperature, the so-called shaking temperature, that would have the same disordering effect. It is proportional to the strength of the disorder present, and inversely proportional to the driving force. Thus, if the disorder in a material is sufficiently strong, the effective shaking temperature may become large enough to melt the vortex lattice. This dynamic melting, the transition from the coherent motion of the vortex lattice to the plastic dynamics of amorphous vortex structures, has been observed experimentally in superconducting films. The effect is also very general: In particular, the effect of abrupt switching in the dynamics of charge density waves is one manifestation of the transition between the plastic and coherent motions of the vortex lattice.

NONSTANDARD EFFECTS

Although vortex physics governs many aspects of superconductivity, it has some manifestation in HTS materials (particularly those that are ceramics), which, because they have a layered structure, have no analogues in conventional superconductors. In the layered superconductors, the vortices induced by the penetrating magnetic field actually appear as sets of aligned vortex discs ("pancake" vortices). Discs in adjacent layers couple via the Josephson interaction; discs in more remote layers interact magnetically. At high magnetic fields, the vortex system becomes so dense that interactions between pancakes within the same layers are stronger than interlayer interactions, and a decoupling transition occurs. The vortex system becomes a "gas" of pancake vortices. (A similar effect can be caused by introducing disorder into the system. Hypothetically, melting the vortex lattice would give rise to a loss of coherence along the magnetic field lines.) The resulting gas of pancakes behaves like a gas of charged particles, with Josephson interactions between pancakes in adjacent layers taking the place of Coulomb interactions between particles. This state of liquid vortex matter is called a Josephson plasma.

Electromagnetic excitations, called plasmons, have been found in Josephson plasmas just as in normal plasmas, as have the usual plasma resonant absorptions, which are called Josephson plasma resonances. Josephson plasma resonance is one of the main diagnostics of the vortex liquid state because the frequency of these resonances directly measures the degree of correlations between the layers. The study of Josephson plasmas has become a major focus of superconductivity research in Japan, motivated in part by its potential technological application. Interactions between a moving Josephson plasma and the underlying periodic potential of the lattice of the material in which it exists may make it possible to generate sharp signals at terahertz frequencies. Devices exploiting these effects might include high-resolution filters, high-sensitivity detectors, and AC field generators in the terahertz frequency range (for terahertz lasers).